THINK PHYSICS

BEGINNER'S GUIDE TO AN AMAZINGLY WIDE RANGE
OF FUNDAMENTAL PHYSICS- RELATED QUESTIONS

THINK
PHYSICS
BALUNGI FRANCIS

PREFACE

Balungi explains deep ideas in physics in an easy-to-understand way. Think Physics is a series aimed to solving the big problems in physics. The book targets topics that researchers and students spend time wondering about, like the origin of gravity and the universe. It also goes into the theories that seem right but are wrong and shows why they are wrong a rarity in science books. Think Physics series is a rigorously correct, lighthearted, and cleverly designed problem solving book for physicists of all ages.

? Has been tested, rewritten, and retested to ensure that you can teach yourself all about major unsolved physics problems

? Requires no math-mathematical treatments and applications are included in optional sections so that you can choose either a mathematical or nonmathematical approach. **No Calculus**

The Minimum mass limit of a gravitationally Collapsed star

The smallest black hole would be one where the Schwarzschild radius equals the radius of a mass with a reduced Compton wavelength which is the smallest size to which a given mass can be localized. For a small mass M, the Compton wavelength exceeds half the Schwarzschild radius, and no black hole description exists. This smallest mass for a black hole is thus approximately the Planck mass, the micro black hole.

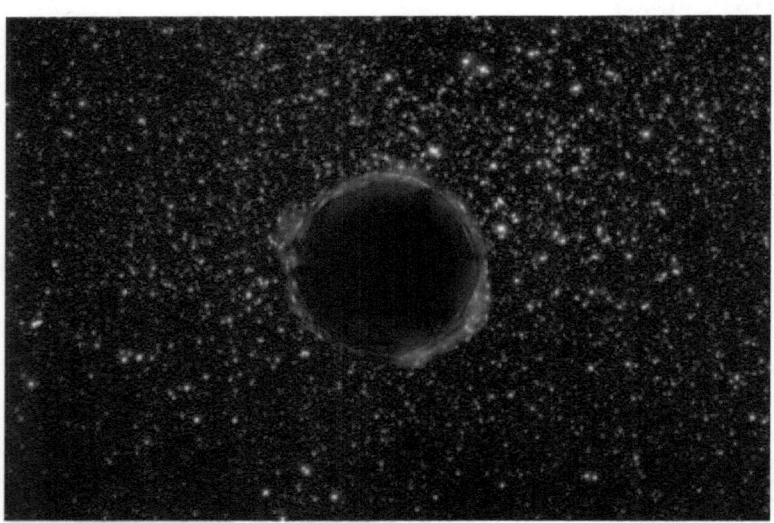

Contrary to the above observation, torsion (see Einstein-Cartan theory) modifies the Dirac equation in the presence of the gravitational field causing fermions to be spatially extended. This spatial extension of fermions limits the minimum mass of a black hole to be on the order of $10^{16} Kg$, showing that micro black holes (of Planck mass) may not exist. Another mass limit is from the data of the Fermi Gamma-ray space telescope satellite which states that, less than one percent of dark matter could be made of primordial black holes with masses up to $10^{13}Kg$.

The major aim of this book is to prove theoretically the existence of a minimum mass limit of a gravitationally collapsed star and thereafter prove **Chandrasekhar** wrong (see Chandrasekhar 1983 Noble lecture concluding statement below)

"We conclude that there can be no surprises in the evolution of stars of mass less than 0.43Solarmass ($\mu = 2$). The end stage in the evolution of such stars can only be that of the white dwarfs. (Parenthetically, we may note here that the so-called 'mini' black-holes of mass 10^{12} Kg cannot naturally be formed in the present astronomical universe.)"

In what follows the above given statement may prove to wrong according to a detailed derivation given below.

From the theory of white dwarf stars, the radius limit of a white dwarf of mass M is given by the following equation,

$$R_w = \frac{(9\pi)^{2/3} \hbar^2}{8} \frac{1}{m_e G (m_{pro})^{5/3} M^{1/3}}$$

$$(1)$$

Where m_{pro} and m_e is the proton and electron mass respectively

Just like the Compton wavelength, there must exist another radius for the consistitution of stars that differs from the radius given in (1) above. For example, in the same way the Planck mass is deduced (i.e by equating the Schwarzschild radius to the Compton wavelength) is the same way in which we are to prove the existence of the mass limit of a gravitationally collapsed star.

We start from first principles. Let it be known that the derivation of the Chandrasekhar mass limit will follow the equipartition of the gravitational potential energy of a star to its electron degeneracy pressure. In the same way, if the gravitational binding energy is given by,

$$E_g = \frac{2M_{pl}{}^3 m_e c^2 (6.144\pi^3)}{M m_{pro}{}^2 \quad \mu^2} \qquad (2)$$

Where M_{pl} is the Planck mass and μ is is the average molecular weight per electron

And the electron degeneracy energy pressure of the star is given by,

$$E_d = m_e c^2$$

When $E_q = E_d$ then we obtain the mass limit of the white dwarf star as,

$$M = \frac{12.288\pi^3 M_{pl}{}^3}{\mu_e{}^2 \quad m_p{}^2} = 1.4 M_{sun}$$

If then this is true, then the formula (2) for the gravitational binding energy of a star is true. This therefore implies that the following assumption will also be true.

When the binding gravitational energy of a star is equal to the Newtonian gravitational potential energy $\frac{GM^2}{R}$ we obtain the radius which is the smallest size to which a given mass of a star can be localized as,

$$\frac{GM^2}{R} = \frac{2M_{pl}^3 m_e c^2 (6.144\pi^3)}{M m_{pro}^2 \quad \mu^2}$$

$$R = \frac{G m_{pro}^2}{2 m_e c^2}\left(M/M_{pl}\right)^3 \frac{\mu^2}{6.144\pi^3}$$

$$(3)$$

This can be rewritten in the form,

$$R = R_k \left(M/M_{pl}\right)^3$$

Where $R_k = 2.384 \times 10^{-53}m$ which is smaller than the Planck length of $1.62 \times 10^{-35}m$

Therefore equating Equation (1) to Equation (3) we deduce the mass limit of a gravitationally collapsed star as,

$$M = \left(293.534\pi^{11}M_{pl}^{21}/\mu^6 M_{pro}^{11}\right)^{1/10} = 9.54 \times 10^{13}Kg$$

The value is in excellent agreement with other theoretical and experimental observations

The radius of this black hole from Equation (3) is thus $2.527 \times 10^{14} m$ larger than the radius of the sun of $7 \times 10^{8} m$

.

In conclusion therefore the end stage in the evolution of a star can only be that of the black hole with a mass $9.54 \times 10^{13} Kg$ and size of $2.53 \times 10^{14} m$ in contrast with the Chandrasekhar observations.

Note that the radius given by Equation (3), $R = R_k \left({}^{M}\!/\!_{M_{pl}} \right)^3$ above is similar to the Equation for the size of the Planck star that was given by Rovelli and Vidotto, $r = l_p \left(\frac{M}{M_{pl}} \right)^n$ where l_p is the Planck length and n is the positive number. This is a clear indication that singularities in black holes can be resolved.

Emergence of Gravity

Starting from first principles and general assumptions we present a heuristic argument that shows that Newton's law of gravitation naturally arise in a theory in which space emerges through a zero- point fluctuation of the quantum vacuum. Gravity is identified with a casimir force caused by quantum vacuum fluctuations due to the presence of material bodies in it or the distortion of the vacuum through its interaction with mass. A relativistic generalization of the presented arguments directly leads to the Einstein equations. When space is emergent even Newton's law of inertia needs to be explained. The equivalence principle suggests that it is actually the law of inertia whose origin is casimir.

The real origin of gravity is one of the most important, complex and substantially yet unsolved questions in Physics. The replacement of the Newtonian model of gravity with the Einstein's one given by General Relativity (GR) has only shifted the question without solving it. Within GR, gravity has two possible interpretations: a field one and a geometric one. According to the latter, that has become the prevalent one, gravity is due to the curvature of the space – time "tissue", represented as a "rubber sheet", due to the presence of a mass. Nevertheless, this is a purely mathematical description telling nothing about the physical mechanism starting the motion. In fact, even supposing the existence, in the neighbouring of a source mass, of a curved four dimensional manifold it doesn't explain why a second particle at rest should move towards the source mass.

As such, it invites attempts at derivation from a more fundamental set of underlying assumptions, and six such attempts are outlined in the standard reference book Gravitation, by Misner, Thorne, and Wheeler (MTW). ' Of the six approaches presented in MTW, perhaps the most

9

far-reaching in its implications for an underlying model is one due to Sakharov; namely, that gravitation is not a fundamental interaction at all, but rather an induced effect brought about by changes in the quantum fluctuation energy of the vacuum when matter is present. 'In this view the attractive gravitational force is more akin to the induced van der Waals and Casimir forces, than to the fundamental Coulomb force. Although speculative when first introduced by Sakharov in 1967, this hypothesis has led to a rich and ongoing literature on quantum-fluctuation-induced gravity that continues to be of interest.

Many physicists believe that gravity, and space-time geometry are emergent. Also string theory and its related developments have given several indications in this direction.

In this book we will argue that the central notion needed to derive gravity is vacuum polarization. More precisely, the presence of matter in the vacuum is taken to constitute a kind of set of boundaries as in a generalized Casimir effect, and the question of how quantum fluctuations of the vacuum under these circumstances can lead to an action and metric that reproduce Einstein gravity will be addressed from several viewpoints. **We want to show that gravitation might be not a fundamental interaction but a byproduct of the electromagnetic interaction, precisely an electromagnetic phenomena induced by the presence of matter in the quantum vacuum (the quantum field that is present even in empty space).** Which means that, matter is not just there but is in the quantum vacuum, and therefore interacts with it, causing some kind of quantum fluctuation energy, that fluctuation is gravitation. In simple terms, a body immersed in quantum fields will interact with them causing gravity to manifest.

Changes in this vacuum when matter is displaced leads to a reaction force. Our aim is to show that this force, given certain reasonable assumptions, takes the form of gravity.

Vacuum polarization describes a process in which a background electromagnetic field produces virtual electron-positron pairs that change the distribution of charges and currents that generated the original electromagnetic field.

The effects of Vacuum polarization have been observed experimentally, for example in measurements of the lamb shift and anomalous magnetic dipole moment of the electron for a precise determination of the fine structure constant. These effects were calculated to first order in the coupling constant by R. serber and E.A Uehling in 1935.

The law of gravitation is derived from classical Casimir effect, which states that the vacuum through which particles move is not empty but consists of an indeterminate state of fluctuating fields and particles.

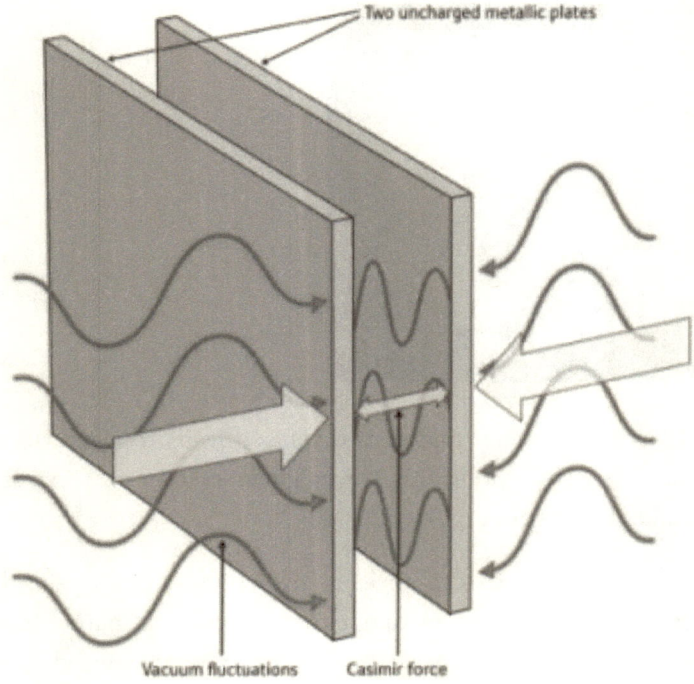

Two uncharged metallic plates

Vacuum fluctuations   Casimir force

Even if you remove all the particles and radiation from a region of space, space still won't be empty. It will consist of virtual pairs of particles and antiparticles.

It is known that, when light propagates through an "empty" region, if space is perfectly empty, it should move through that space unimpeded, without bending, slowing or breaking into multiple wavelengths. Applying an external magnetic field doesn't change this, as photons, with their oscillatory electric and magnetic fields don't bend in a magnetic field. Even when your space is filled with particle antiparticle pairs, this effect doesn't change. But if you apply a **strong magnetic field** to a space filled with particle antiparticle pairs, suddenly a real, observable effect arises.

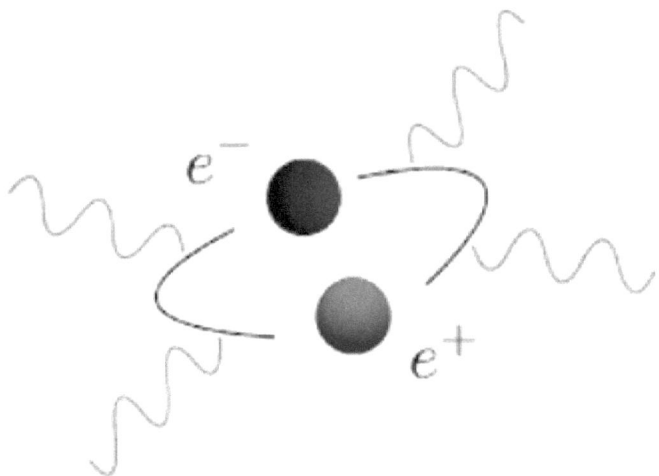

A virtual particle is a transient quantum fluctuation that exhibits some of the characteristics of an ordinary particle, while having its existence limited by the uncertainty principle. The longer the virtual particle exists, the closer its characteristics come to those of ordinary particles. They are important in the physics of many processes, including particle scattering and Casimir forces.

In 2015 theoretical physicist James Quach, suggested a way to detect gravitons by taking advantage of their quantum nature. Quantum mechanics suggests the universe is inherently fuzzy for instance, one can never absolutely know a particle's position and momentum at the same time. One consequence of this uncertainty is that a vacuum is never completely empty, but instead buzzes with a "quantum foam" of so-called virtual particles that constantly pop in and out of existence. These ghostly entities may be any kind of quanta, including gravitons.

Decades ago, scientists found that virtual particles can generate detectable forces. For example, the Casimir effect is the attraction or repulsion seen between two mirrors placed close together in vacuum. These reflective surfaces move due to the force generated by virtual photons winking in and out of existence. Previous research suggested that superconductors might reflect gravitons more strongly than normal matter, so Quach calculated that looking for interactions between two thin superconducting sheets in vacuum could reveal a gravitational Casimir effect. The resulting force could be roughly 10 times stronger than that expected from the standard virtual-photon-based Casimir effect.

In quantum field theory, even classical forces-such as the electromagnetic repulsion or attraction between two charges-can be thought of as due to the exchange of many virtual photons between the charges. Virtual photons are the exchange particle for the electromagnetic interaction.

From the energy time uncertainty principle, we deduce the Casimir force in this way,

$$\Delta E \Delta t \sim \hbar$$

Considering two electrons separated by a distance R in a vacuum, the work done by a photon to move from one charge to the other is equivalent to the product of the force F applied and the distance R through which it moves

$$\Delta E = F \Delta R$$

Therefore the force carried by the virtual photon (messenger particles) between the two charges is

$$F = \frac{\hbar}{\Delta R \Delta t}$$

But since the virtual photons move at a constant speed of light c in the vacuum between the electrons separated by a distance ΔR, inserting

$$c = \frac{\Delta R}{\Delta t}$$

The Casimir force is given as,

$$F = \frac{\hbar c}{R^2}$$

We notice that the force deduced above is similar to the gravitational force because it falls off as the inverse of the distance squared.

An electron feels a force like gravity due to the changes in the vacuum (exchange of gravitons) brought about by the presence of a strong magnetic field.

Recently, Norte and his colleagues developed a microchip to perform this experiment. This chip held two microscopic aluminum-coated plates that were cooled almost to absolute zero so that they became superconducting. One plate was attached to a movable mirror, and a laser was fired at that mirror. If the plates moved because of a gravitational Casimir effect, the frequency of light reflecting off the mirror would measurably shift. However the scientists failed to see any gravitational Casimir effect. This null result does not necessarily rule out the existence of gravitons and thus gravity's quantum nature. Rather, it may simply mean that gravitons do not interact with superconductors as strongly as prior work estimated, says quantum physicist and Nobel laureate Frank Wilczek of the Massachusetts Institute of Technology, who did not participate in this study and was unsurprised by its null results. Even so, Quach says, this "was a courageous attempt to detect gravitons."

Although Norte's microchip did not discover whether gravity is quantum, other scientists are pursuing a variety of approaches to find gravitational quantum effects. For example, in 2017 two independent studies suggested that if gravity is quantum it could generate a link known as "entanglement" between particles, so that one particle instantaneously influences another no matter where either is located in the cosmos. A table top experiment using laser beams and microscopic diamonds might help search for such gravity-based entanglement. The crystals would be kept in a vacuum to avoid collisions with atoms, so they would interact with one another through gravity alone. Scientists would let these diamonds fall at the same time, and if gravity is quantum the gravitational pull each crystal exerts on the other could entangle them together.

The researchers would seek out entanglement by shining lasers into each diamond's heart after the drop. If particles in the crystals' centers spin one way, they would fluoresce, but they would not if they spin the other way. If the spins in both crystals are in sync more often than chance would predict, this would suggest entanglement. "Experimentalists all over the world are curious to take the challenge up," says quantum gravity researcher Anupam Mazumdar of the University of Groningen in the Netherlands, co-author of one of the entanglement studies.

To prove that gravity is Casimir, we propose that the virtual photons exchanged between electrons take the force of electromagnetic waves or radiations propagating in the vacuum between the electrons with a changing electric and magnetic field, such that their speed depends on the electric field E and magnetic B as,

$$c = \frac{E}{B}$$

The force felt by an electron due to vacuum polarization then becomes,

$$F = \frac{\hbar E}{BR^2}$$

It was previously deduced (see quantum gravity in a Nutshell1) that the characteristic value of the electric field built from the electron mass is

$$E = \frac{m^2 e c^2}{4\pi\varepsilon_0 \hbar^2} = 9.667 \times 10^{15} N/C$$

Where m is the mass of the electrons, e is the elementary charge and ε_0 is the permittivity of the vacuum between the electrons

This electric field can be observed near Neutron stars or magnetars.

$$F = \frac{m^2 e c^2}{4\pi\varepsilon_0 \hbar BR^2}$$

Suppose we apply a strong external Magnetic field to the vacuum of

$$B = \frac{e c^2}{4\pi\varepsilon_0 G \hbar} = 1.8423 \times 10^{48} T$$

Note: The above given magnetic field was previously derived in my book "quantum gravity in a nutshell1"

We obtain a familiar law between the two electrons

$$F = G\frac{m^2}{r^2}$$

We have recovered Newton's law of gravitation, practically form first principles. Following the above derivation carefully, it implies that gravity is a force resulting from the quantum vacuum polarizations due to an existence of an external strong magnetic field. This is proof that gravity is indeed a quantum force which can be explained from the Casimir effect.

The effects of the given derivation can be experimentally verified near Neutron stars and Magnetars with strong magnetic fields. When the given values of the electric and magnetic fields are observed with high tech experiments then it will prove without doubt that the force felt by the electron in the vaccum is indeed a quantum force.

The above given theoretical model can be verified experimentally from the effect known as Vacuum Birefringence, occurring when charged particles, get yanked in opposite directions by strong magnetic field lines.

The effect of this vacuum birefringence gets stronger very quickly as the magnetic field strength increases, as the square of the field strength. Even though the effect is small, we have place in the universe where the magnetic field strength get large enough to make these effects relevant. This place is the Neutron star and Magnetars with strong magnetic fields.

The outer 10% of a neutron star consists mostly of protons, light nuclei, and electrons, which can stably exist without being crushed at the neutron star's surface.

Neutron stars rotate extremely rapidly, frequently in excess of the 10% the speed of light, meaning that these charged particles on the outskirts of the neutron star are always in motion, which necessitates the production of both electric currents and induced magnetic fields. These are the fields we should be looking for if we want to observe vacuum birefringence, and its effects on the polarization of light.

All the light that's emitted must pass through the strong magnetic field around the neutron star on its way to our eyes, telescope and detectors, if the magnetized space that it passes through exhibits the expected vacuum birefringence effect, that light should all be polarized, with a common direction of polarization for all the photons.

The Simple Link between Quantum Mechanics and Gravity

Today we are blessed with two extraordinarily successful theories of physics. The first is the General theory of relativity, which describe the large scale behavior of matter in a curved space time. This theory is the basis for the standard model of big bang cosmology. The discovery of gravitational waves at LIGO observatory in the US (and then Virgo, in Italy) is only the most recent of this theory's many triumphs.

The second is quantum mechanics. This theory describes the properties and behavior of matter and radiation at their smallest scales. It is the basis for the standard model of particle physics, which builds up all the visible constituents of the universe out of collections of quarks, electrons and force-carrying particles such as photons. The discovery of the Higgs boson at CERN in Geneva is only the most recent of this theory's many truimphs.

But, while they are both highly successful, those two structures leave a lot of important questions unanswered. They are also based on two different interpretations of space and time, and are therefore fundamentally incompatible. We have two descriptions but, as far as we know, we've only ever had one universe. What we need is a quantum theory of gravity.

One of the problems facing physicists is to link Gravity with quantum mechanics in the study of the quantum effects of Black holes. Within the past few years there have been developments that give rise to the hope that before too long we shall have a fully consistent quantum theory of gravity, one that will agree with general relativity for macroscopic objects and will, one hopes, be free of the mathematical infinities that have long bedeviled other quantum field theories. Such developments include; Loop Quantum Gravity and String theory.

These developments have to do with certain recently discovered quantum effects associated with black holes, which provide a remarkable connection between black holes and the laws of thermodynamics.

One of the first attempts towards the unification of quantum mechanics with gravity was made in part by Stephen Hawking and Bekenstein in the late 70s. The black hole creates and emits particles and radiation just as if it were an ordinary hot body with a temperature that is proportional to the surface gravity and inversely proportional to the mass. This made Bekenstein's suggestion that a black hole had a finite entropy fully consistent, since it implied that a black hole could be in

thermal equilibrium at some finite temperature other than zero.

Hawking beautiful result raises a number of questions. First, in Hawking's derivation the quantum properties of gravity are neglected. Are these going to affect the result? Second, we understand macroscopical entropy in statistical mechanical terms as an effect of the microscopical degrees of freedom. What are the microscopical degrees of freedom responsible for the entropy? Can we derive the Bekenstein entropy from first principles? Because of the appearance of $\hbar$ in the entropy formula, it is clear that the answer to these questions has since become a standard benchmark against which a quantum theory of gravity can be tested.

A definitive resolution of the quantum gravity problem will require deriving the Bekenstein-Hawking area-entropy law from first principles without the introduction of adhoc principles.

$$S = \frac{Ac^3k}{4\hbar G}$$

where A is the surface area of the Schwarzschild black hole, c is a constant speed of light, k the Boltzmann constant, $\hbar$ the reduced Planck constant and G is the Newton's gravitational constant.

Attempts towards this were done in the early 70s by Hawking who proved that a black hole emits thermal radiation with a temperature

$$T = \frac{\hbar c^3}{8\pi Gk}$$

This book presents a simple universal explanation of Black hole thermodynamics in a somewhat different form than that given by Loop Quantum Gravity (LQG), String theory and Hawking radiation theory. The major result of the book is the derivation of the temperature and entropy of a black from first principles with a well defined calculation where no infinities appear.

Quantizing Gravity

Gravity is the weakest of the four fundamental forces of physics, approximately 10^{38} times weaker than the strong force, 10^{36} times weaker than the electromagnetic force and 10^{29} times weaker than the weak force. As a consequence, it has no significant influence at the level of subatomic particles. This becomes a problem in reconciling gravity with quantum mechanics. To unit gravity with the other three forces of physics, we propose that gravity be quantized in the following form

$$F_G = \alpha F_I \qquad (1)$$

Where F_G is the gravitational force, α is the quantum number and F_I is one of the other three forces of nature.

In case one wants to determine the strength of the gravitational force in comparison to the other three forces, then one sets α as the gravitational coupling constant which determines the strength of the force

$$\alpha = \frac{F_G}{F_I}$$

Although gravity is one of the weakest forces known in the universe, it however becomes stronger near space time singularities and from Einstein general relativity, a particle at the event horizon of a Black hole (that is at a Schwarzichild radius $R = 2GM/c^2$ from the center of a Black hole) feels a strong force given as

$$F_G = \frac{GM^2}{\pi R^2} = \frac{c^4}{4\pi G}$$

Note from above, the known gravitational force law is inversely proportional to the area of the event horizon of a Black hole not it's radius.

If a black hole was an oscillating particle like an electron, then according to quantum mechanics (from the uncertainty principle), its position will be determined by hitting it with a particle the same size as its mass.

$$x = \frac{\hbar}{2Mc}$$

Then the gravitational coupling constant becomes the ratio of the Schwarzschild radius to the position x

$$\alpha = \frac{R}{x} = \frac{4GM^2}{\hbar c}$$

Inserting in (1) we have a quantum mechanical description of gravity in the form

$$F_G x = R F_I$$

Then the work done by one of the three forces to move a particle from the center of a Black hole (in the form of radiation) to the event horizon through a Schwarzschild radius R against the strong gravitational force is then calculated to be

$$F_I R = \frac{\hbar c^3}{8\pi GM}$$

By the equipartition rule, the thermal energy of a Heat bath of a Black hole at temperature T is hereby equal to the work done by a force F_I to pull a particle from the black hole.

$$kT = F_I R = \frac{\hbar c^3}{8\pi GM}$$

Where k is the Boltzmann constant

Finally we get a familiar temperature

$$T = \frac{\hbar c^3}{8\pi GMk} \qquad (2)$$

We have therefore recovered Hawking temperature from first principles. This means that, the black hole creates and emits particles and radiation just as if it were an ordinary hot body with a temperature that is proportional to the surface gravity and inversely proportional to the mass.

When a Black hole evaporates or emits radiations, it losses mass in time t, the intensity of the emitted thermal or electromagnetic radiations is given by

$$I = \frac{F_I^2}{2\pi\hbar}$$

Where the intensity by definition is power P per unit area A of the event horizon of a black hole

$$\frac{P}{A} = \frac{F_I^2}{2\pi\hbar}$$

The power is the energy E lost by a black hole in time t (The time taken by a Black hole to dissipate)

$$t = \frac{Mc}{F_I}$$

From which the total energy lost by a Black hole is given by

$$E = \frac{AMcF_I}{2\pi\hbar}$$

But since from (1),

$$F_G = \alpha F_I$$

Then

$$E = \frac{AMcF_G}{2\pi\alpha\hbar}$$

Inserting $F_G = \frac{c^4}{4\pi G}$ and $\alpha = \frac{4GM^2}{\hbar c}$, we get

$$E = \frac{Ac^6}{32\pi G^2 M}$$

We have recovered the Frodden-Ghosh- Perez Energy.

Hawking knew that if the horizon area were an actual entropy, black holes would have to radiate. When heat is added to a thermal system, the change in entropy is the increase in thermal energy divided by temperature:

$$S = \frac{E}{T}$$

From (2), the temperature of a black hole is given by,

$$T = \frac{\hbar c^3}{8\pi GMk}$$

Therefore the Entropy of a Black hole will be given by,

$$S = \frac{\left(\dfrac{Ac^6}{32\pi G^2 M}\right)}{\left(\dfrac{\hbar c^3}{8\pi GMk}\right)} = \frac{Akc^3}{4G\hbar}$$

This implies that the energy radiated by a black hole goes into increasing its event horizon area and an increase in area will automatically lead to an increase in entropy as explained by the above given area entropy law. As far as this book is concerned there is no other theory from which such a calculation can proceed. Hence the book is the only one from which a detailed quantum theory of gravity precedes and where the result of the Bekenstein-Hawking area entropy law can be achieved.

Proof of the Proton Radius

Today the proton radius is measured via three methods that is, the spectroscopy, nuclear scattering and muonic hydrogen (2010 experiment) methods.

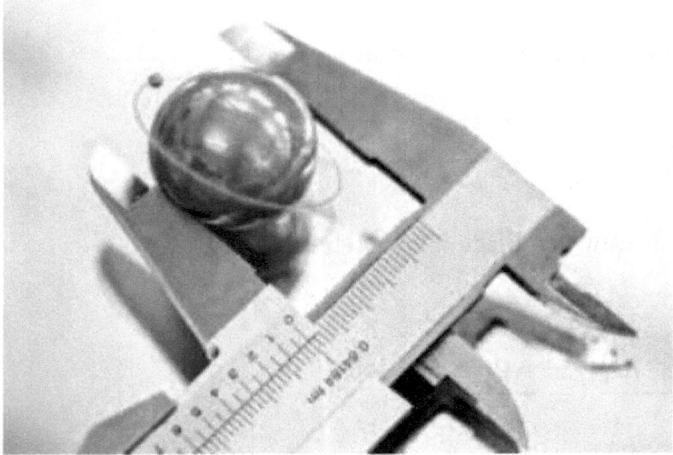

The spectroscopy method uses the energy levels of electrons orbiting the nucleus. This method produces a proton radius of about 8.768×10^{-16} m, with approximately 1% relative uncertainty.

The nuclear method is similar to Rutherford's scattering experiments that established the existence of the nucleus. Small particles such as electrons can be fired at a proton, and by measuring how the electrons are scattered, the size of the proton can be inferred. Consistent with the spectroscopy method, this produces a proton radius of about 8.775×10^{-16} m.

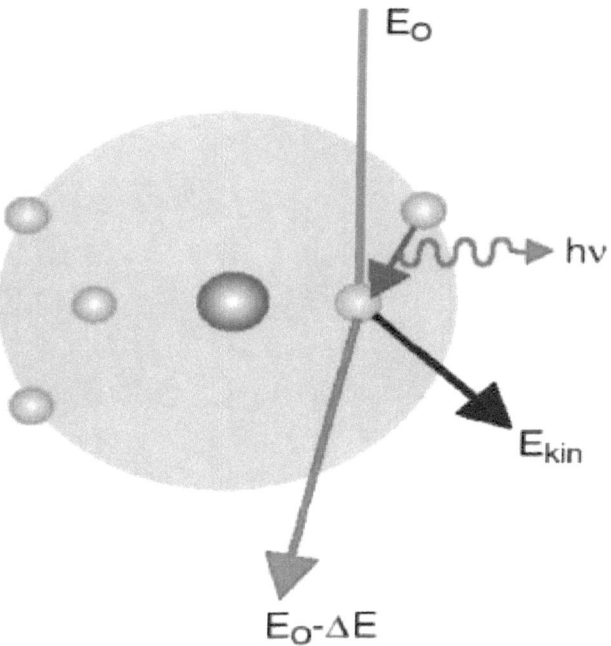

The muonic hydrogen 2010 method by Pohl et al. is similar to the spectroscopy method. However, the much higher mass of a muon causes it to orbit 207 times closer than an electron to the hydrogen nucleus, where it is consequently much more sensitive to the size of the proton. The resulting radius was recorded as 8.42×10^{-16} m. This newly measured radius is 4% smaller than the prior measurements, which were believed to be accurate within 1%.

The discrepancy between the measured values of the proton radius by the methods given above is what is called the proton radius puzzle and the discrepancy might be due to new physics, or the explanation may be an ordinary physics effect that has been missed.

In what follows, I deduce the radius of the proton from first principles using a new approach.

Consider two protons of mass m_{pro} at a distance R apart in an **antiprotonic hydroden atom** . Let the protons be circular with area A.

To measure the radius of the proton, we consider a **protonium**, also known as **antiprotonic hydroden**, in which a proton and an antiproton orbit each other.

According to the Heisenberg uncertainty principle, the space between the protons is not empty. Therefore the calculated energy density contained in the vacuum between the protons is then related to the Casimir force as

$$\rho_{cal} = \frac{F_c}{A} = \frac{2\hbar c}{AR^2} \qquad (1)$$

Where A is the area of the proton, $\hbar$ is the reduced Planck constant and c is the constant speed of light.

However, from general assumptions not given here, the observed/ experimental vacuum energy density is proportional to the square of the force in any given interaction as,

$$\rho_{obs} = \frac{F^2}{8\pi\alpha\hbar c} \qquad (2)$$

Where α is the coupling constant (which determines the strength of the force in any given interaction)

At a point where the calculated vacuum energy density due to the Casimir effect is equal to the observed vacuum energy density due to the known fundamental force between the protons, the proton will move in a circular orbit around the other proton. The force of attraction between the two protons will then be deduced (by equating (1) to (2))

$$F = \frac{4\hbar c}{R}\sqrt{\frac{\pi\alpha}{A}} \qquad (3)$$

As the proton orbits closer to the other proton, its speed increases until it comes into contact with the other proton. Suppose the proton is moving at a constant speed of light c in its orbit, then the centripetal force that keeps it in orbit is equal to the force of attraction between them.

$$\frac{m_{pro}c^2}{R} = \frac{4\hbar c}{R}\sqrt{\frac{\pi\alpha}{A}}$$

Arranging and cancelling like terms, we get the formula for the area of the proton as

$$A = \frac{16\pi\alpha\hbar^2}{m_{pro}^2 c^2}$$

Inserting, $A = \pi r_{pro}^2$, where r_{pro} is the proton radius

$$r_{pro} = \frac{4\hbar}{m_{pro}c}\sqrt{\alpha}$$

The force required to separate two protons apart is indeed the strong nuclear force whose strength is determined by the strong coupling constant $\alpha = 1$, from which we calculate the proton radius as,

$$r_{pro} = \frac{4\hbar}{m_{pro}c} = 8.4075 \times 10^{-16}m$$

This value is very close to the 2010 Pohl result of an experiment relying on muonic hydrogen, practically from first principles! See Paul Scherrer institute in Switzerland (CREMA-Charge Radius Experiment with Muonic Atoms) vol466/8 july2010/doi:10.1038/ nature09250

Reinventing Gravity

Just as Einstein in his day was constructing an alternative gravity theory, an alternative to Newton's law of gravity,

which had prevailed for more than 200 years, so I have been searching for a larger theory, a modification of general relativity that would fit the data without the need to posit dark matter, and would contain Einstein's theory

just as Einstein contains Newton. Unlike the alternative gravity theories known today, my mature Modified Gravity theory, contains no physical instabilities. It is as robust a gravity theory as general relativity, and fits all the current astrophysical and cosmological data without dark matter.

Modified Newtonian Dynamics (MOND) by Milogram predicted a modified Newtonian acceleration law that could fit the large amount of anomalous rotational velocity curve data from galaxies obtained by astronomers since the late 1970s, which showed stars rotating at the edges of galaxies at twice the rate predicted by Newton and Einstein. My aim is to try to explain the data without the conventional reliance on exotic dark matter.

Background

In Newton's view, all objects exert a force that attracts other objects. That universal law of gravitation worked pretty well for predicting the motion of planets as well as objects on Earth and it's still used, for example, when making the calculations for a rocket launch. But Newton's view of gravity didn't work for some things, like Mercury's peculiar orbit around the sun. The orbits of planets shift over time, and Mercury's orbit shifted faster than Newton predicted. In spiral galaxies, the orbiting of stars around their centers seems to strongly disobey Newton's law of universal gravitation.

During the 1970s, astronomers discovered something odd about the movement of stars in galaxies. Like the planets orbiting our sun, the stars should follow Newton's law of gravity, and travel ever more slowly the further out they are from the galactic centre. Yet beyond a certain distance, their speeds remained more or less constant - in flat contradiction of Newton's law.

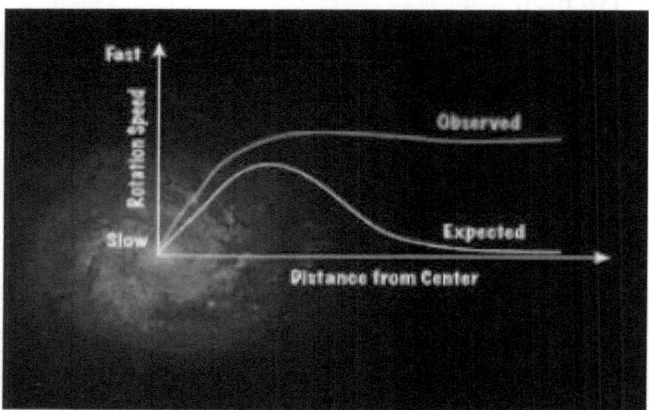

Astronomers quickly proposed a solution: that there are huge amounts of invisible "dark matter" lurking in and around galaxies, whose gravitational pull invisibly affects the stars. But Prof Milgrom had a more radical proposal: that there is something wrong with the law of gravity itself. His calculations suggested that the anomalous motion of the stars could be explained if Newton's law breaks down for masses accelerating below a critical rate of around one ten-billionth of a metre per second per second.

For over 25 years Professor Mordehai Milgrom of the Weizmann Institute in Israel has been pursuing the possibility that both Newton and Einstein missed something when they devised their theories of this most ubiquitous of forces.

According to Einstein, mass warps the very fabric of space and time around it, rather like a cannonball sitting on a vast rubber sheet. This creates the illusion that objects moving past some mass are accelerated by a mysterious "force" emanating from it. In reality, they are just responding to the distortion of space and time - the effect of which is described in detail by Einstein's theory, and captured pretty well even by Newton's simple formula.

Since the early 1980s, Prof Milgrom has suspected there is another flaw in Newton's venerable formula - one which even Einstein failed to fix. And after decades of being ignored by the scientific establishment, there is mounting evidence that he is right.

Prof Milgrom's theory goes by the prosaic name of Modified Newtonian Dynamics or MOND, and is based the bizarre idea that Newton's law of gravity breaks down at low accelerations. And he means very low: around 100-billionth that generated by the Earth's gravity. Like Newton, Prof Milgrom was inspired by a simple observation - albeit a rather more esoteric one than the fall of an apple.

Invention

MOND was proposed by Mordehai Milgrom in 1983. The basic premise of MOND is that while Newton's laws have been extensively tested in high-acceleration environments, they have not been verified for objects with extremely low acceleration, such as stars in the outer parts of galaxies. Several independent observations point to the fact that the visible mass in galaxies and galaxy clusters is insufficient to account for their dynamics, when analysed using Newton's laws.

While Newton's Laws predict that stellar rotation velocities should decrease with distance from the galactic centre, Rubin and collaborators found instead that they remain almost constant the rotation curves are said to be "flat". This observation necessitates at least one of the following:

1) There exists in galaxies large quantities of unseen matter which boosts the stars' velocities beyond what would be expected on the basis of the visible mass alone, or

2) Newton's Laws do not apply to galaxies. The former leads to the dark matter hypothesis; the latter leads to MOND.

In the disc galaxies most of the mass is at the centre of the galaxy, this means that if you want to calculate how a star moves far away from the centre it is a good approximation to only ask what is the gravitational pull that comes from the centre bulge of the galaxy.

Einstein taught us that gravity is really due to the curvature of space and time but in many cases it is still quantitatively incorrect to describe gravity as a force, this is known as the Newtonian limit and is a good approximation as long as the pull of gravity is weak and objects move much slower than the speed of light. It is a bad approximation for example close by the horizon of a black hole but it is a good approximation for the dynamics of galaxies that we are looking at here.

It is then not difficult to calculate the stable orbit of a star far away from the centre of a disc galaxy. For a star to remain on its orbit, the gravitational pull must be balanced by the centrifugal force,

$$\frac{mv^2}{R} = \frac{GMm}{R^2}$$

You can solve this equation for the velocity of the star and this will give you the velocity that is necessary for a star to remain on a stable orbit,

$$v = \sqrt{\frac{GM}{R}}$$

As you can see the velocity drops inversely with the square root of the distance to the centre. But this is not what we observe, what we observe instead is that the velocity continue to increase with distance from the galactic centre and then they become constant.

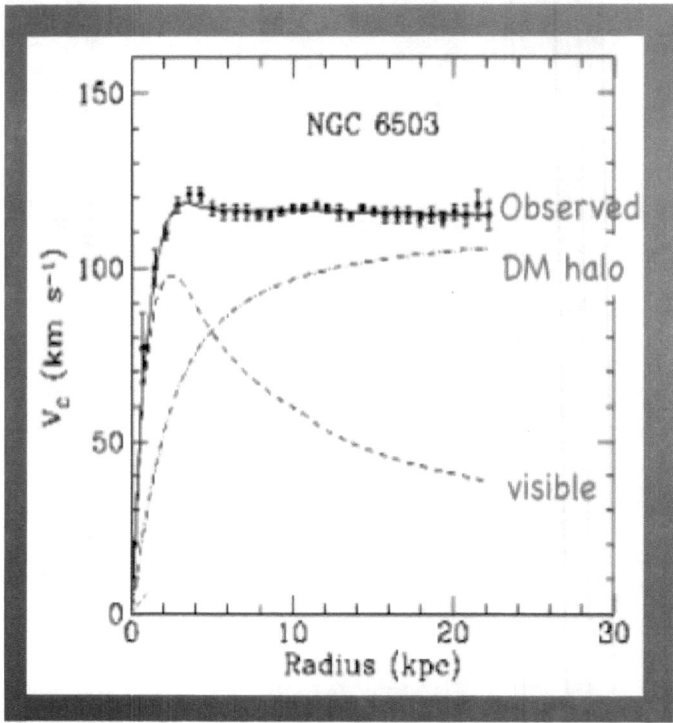

Galaxy data that show that Newtonian and Einstein gravity do not fit the observed speed of stars in orbits inside a galaxy such as NGC 6503

This is known as the flat rotation curve. This is not only the case for our own galaxy but it is the case for hundred of galaxies that have been observed. The curves don't always become perfectly constant sometimes they have rigorous lines but it is abundantly clear that these observations cannot be explained by the normal matter only.

Dark matter solves this problem by postulating that there is additional mass in galaxies distributed in a spherical halo. This has the effect of speeding up the stars because the gravitational pull is now stronger due to the mass from the dark matter halo. There is always a distribution of dark matter that will reproduce whatever velocity curve we observe.

In contrast to this, Modified Newtonian Dynamics (MOND) postulates that gravity works differently. In MOND, the gravitational potential is the logarithmic of the distance

$$\Phi = (\sqrt{GMa_o})\ln\left(\frac{R}{GM}\right)$$

and not as normally the inverse of the distance

$$\Phi = \frac{-GM}{R}$$

In MOND the gravitational force is then the derivation of the potential that is, the inverse of the distance

$$F = \frac{\sqrt{GMa_o}}{R}$$

while normally it is the inverse of the square of the distance

$$F = \frac{GMm}{R^2}$$

If you put this modified gravitational force into the force balance equation as before

$$\frac{\sqrt{GMa_o}}{R} = \frac{v^2}{R}$$

you will see that the dependence on the distance cancels out and the velocity just becomes constant. Now of course you cannot just go and throw out the normal $\frac{1}{R^2}$ gravitational force law because we know that it works on the solar system. Therefore MOND postulates that the normal $\frac{1}{R^2}$ law crosses over into a $\frac{1}{R}$ law. This crossover happens not at a certain distance but it happens at a certain acceleration.

The New force law comes into play at low acceleration a_0, this acceleration where the crossover happens is a free parameter in MOND. You can determine the value of this pararmeter by just trying out which fits the data best. It turns out that the best fit value is closely related to the cosmological constant $a_0 \approx \sqrt{\frac{\Lambda}{3}}$, why does that so? No one has any idea and it is the aim of this section to find out why.

In addition to demonstrating that rotation curves in MOND are flat, it provides a concrete relation between a galaxy's total baryonic mass (the sum of its mass in stars and gas) and its asymptotic rotation velocity. Observationally, this is known as the baryonic Tully Fisher relation and is found to conform quite closely to the MOND prediction.

Milgrom's law fully specifies the rotation curve of a galaxy given only the distribution of its baryonic mass. In particular, MOND predicts a far stronger correlation between features in the baryonic mass distribution and features in the rotation curve than does the dark matter hypothesis.

It predicts a specific relationship between the acceleration of a star at any radius from the centre of a galaxy and the amount of unseen (dark matter) mass within that radius that would be inferred in a Newtonian analysis. This is known as the "mass discrepancy-acceleration relation", and has been measured observationally.

One aspect of the MOND prediction is that the mass of the inferred dark matter go to zero when the stellar centripetal acceleration becomes greater than $a0$, where MOND reverts to Newtonian mechanics. In dark matter hypothesis, it is a challenge to understand why this mass should correlate so closely with acceleration, and why there appears to be a critical acceleration above which dark matter is not required.

In MOND, all gravitationally bound objects with $a < a0$ regardless of their origin – should exhibit a mass discrepancy when analysed using Newtonian mechanics, and should lie on the BTFR. Under the dark matter hypothesis, objects formed from baryonic material ejected during the merger or tidal interaction of two galaxies are expected to be devoid of dark matter and hence show no mass discrepancy. Three objects unambiguously identified as Tidal Dwarf Galaxies appear to have mass discrepancies in close agreement with the MOND prediction.

Recent work has shown that many of the dwarf galaxies around the Milky Way and Andromeda are located preferentially in a single plane and have correlated motions. This suggests that they may have formed during a close encounter with another galaxy and hence be Tidal Dwarf Galaxies. If so, the presence of mass discrepancies in these systems constitutes further evidence for MOND.

By itself, Milgrom's law is not a complete and self-contained physical theory, but rather an ad-hoc empirically motivated variant of one of the several equations that constitute classical mechanics. Its status within a coherent non-relativistic hypothesis of MOND is akin to Kepler's third law within Newtonian mechanics; it provides a succinct description of observational facts, but must itself be explained by more fundamental concepts situated within the underlying hypothesis.

The majority of astrophysicists and cosmologists accept dark matter as the explanation for galactic rotation curves, and are committed to a dark matter solution of the missing-mass problem. MOND, by contrast, is actively studied by only a handful of researchers. The primary difference between supporters of ΛCDM and MOND is in the observations for which they demand a robust, quantitative explanation and those for which they are satisfied with a qualitative account, or are prepared to leave for future work.

This invisible and undetected matter removes any need to modify Newton's and Einstein's gravitational theories. Invoking dark matter is a less radical, less scary alternative for most physicists than inventing a new theory of gravity.

If dark matter is not detected and does not exist, then Einstein's and Newton's gravity theories must be modified. Can this be done successfully? Yes! My Fifth force can explain the astrophysical, astronomical and cosmological data without dark matter.

To differ from the Milogrom approach, we propose that gravity is not a fundamental interaction but rather an emergent property from the vacuum polarizations. In brief gravity is a Casimir effect (see ThinkPhysics2)

From the Uncertainty principle we derive the Casimir force felt by a particle in a vacuum "empty space" (not space is not empty)

$$F = \frac{4\pi\alpha\hbar c}{A}$$

Where α is the coupling constant which determines the strength of the force in an interaction, $\hbar$ is the reduced Planck constant, c is the constant speed of light and A is the surface area.

But since the energy density or pressure of matter confined in a given region of "empty" space is force per unit area or energy per unit volume,

$$\rho = \frac{F}{A}$$

Then the reaction force felt by the space in which a particle rests or moves is given as,

$$F = \sqrt{4\pi\alpha\hbar c\rho}$$

Applying this force to the entire galaxy, Now suppose our boundary is not infinitely extended, but forms a closed surface. More specifically, let us assume it is a sphere with already emerged space on the outside.

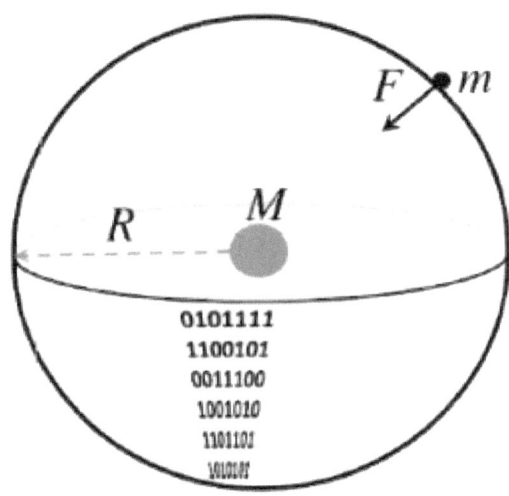

A particle with mass m near a spherical holographic screen. The magnetic field is evenly distributed over the occupied bits, and is equivalent to the mass M that would emerge in the part of space surrounded by the screen.

Let the surface of the boundary be evenly distributed with a positive elementary charge +e creating in turn a negative charge on the inside.

The key statement is simply that we need to have vacuum energy density in order to have a force between the two masses. Since we want to understand the origin of gravity, we need to know where the vacuum energy density comes from.

One can think about the boundary as a storage device for information. Assuming that the holographic principle holds, the maximal storage space, or total number of bits, is proportional to the internal Electric Field created by the masses. In fact, in a theory of emergent space this is how magnetic field may be defined: each fundamental bit occupies by definition one unit cell.

Let us denote the number of used bits by ?. It is natural to assume that this number will be proportional to the Electric field. So we write as before (see ThinkPhysics2),

$$\alpha = \frac{E_0 c}{B} = \frac{\left[\dfrac{Mmec}{4\pi\varepsilon_0 \hbar^2}\right]}{\left[\dfrac{ec^2}{4\pi\varepsilon_0 G\hbar}\right]} = \frac{GMm}{\hbar c}$$

Where we introduced a new constant G. Eventually this constant is going to be identified with Newton's constant, of course. But since we have not assumed anything yet about the existence of a gravitational force, one can simply regard this equation as the definition of G. So, the only assumption made here is that the number of bits is proportional to the characteristic Electric field from the mass. Nothing more.

Suppose there is a total vacuum energy density ρ present in the system. The vacuum energy density is determined by the equipartition rule

$$\rho = \frac{F}{A}$$

Finally one inserts

$$A = 4\pi R^2$$

The energy density is then defined as the force on particles accelerating in the vacuum per unit area,

$$\rho = \frac{F_o}{4\pi R^2}$$

From Newton's second law, the force on the particle is related to the acceleration by,

$$F_o = ma_o$$

In contrast to Dark matter, our fifth force postulates that gravity works differently. The gravitational force is the inverse of the distance

$$F = \frac{1}{R}\sqrt{a\hbar cma_o}$$

If you put this fifth force into the force balance equation as before

$$\frac{mv^2}{R} = \frac{1}{R}\sqrt{\alpha\hbar cma_0}$$

you will see that the dependence on the distance cancels out and the velocity just becomes constant. Finally one inserts the number of bits/gravitational coupling constant

$$\alpha = \frac{GMm}{\hbar c}$$

and one obtains the familiar relation

$$v^4 = a_0 GM$$

We have recovered the Tully-Fisher relation, practically from first principles! It is one of the best fit predictions for MOND " The relation between asymptotic velocity and the mass of the galaxy is an absolute one" (Milgrom 1983). This is given by, $v^4 = a_0 GM$, where $a_0 = 1.2 \times 10^{-10} ms^{-2}$. It is this behavior that gives rise to asymptotically flat rotation curves and the Tully-Fisher relation (Tully & Fisher 1977) without invoking dark matter.

This Fifth force is an alternative to the hypothesis of dark matter in terms of explaining why galaxies do not appear to obey the currently understood laws of physics. It is an alternative to Entropic gravity, MOND, Quantum gravity and General relativity. Now of course you cannot just go and throw out the normal

$$\frac{1}{r^2}$$

gravitational force law because we know that it works on the solar system. Therefore the normal $\frac{1}{r^2}$ law crosses over into a $\frac{1}{r}$ law. This crossover happens not at a certain distance but it happens at certain acceleration. The New force law comes into play at low acceleration $a_0 = 1.2 \times 10^{-10} ms^{-2}$, this acceleration where the crossover happens is a free parameter in MOND. You can determine the value of this pararmeter by just trying out which fits the data best. It turns out that the best fit value is closely related to the cosmological constant

$$a_0 \approx \sqrt{\frac{\Lambda}{3}}$$

We anticipate that this fifth force will modify how stars collapse and the nature of black holes.

On the Deflection of Light in the Sun's Gravitational Fields

The Newtonian Approach

During Newton's time, it was believed that light was made up of particles moving at a varying speed. To prove why light bends near the Sun's surface Newton had to assume that these particles had mass. For example he considered a Sun with mass M, where a particle of light with mass m from a distant star past the Sun, had to bend near the Sun's surface due to the gravitational force of attraction acting on the particle of light. Because of this, the observer at the earth's surface never saw the actual position of the star but rather the apparent position of the star at an angle θ from its original position.

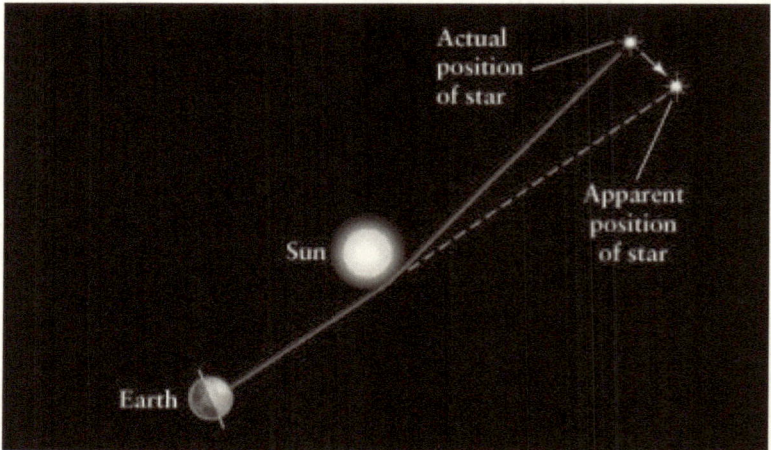

Newton assumed that, the particle of light falling freely in the gravitational field of the Sun gained kinetic energy,

$$E_k = \frac{mv^2}{2}$$

Where, v was the speed of the particle of light.The potential gravitational energy that was gained by the particle was given by,

$$V_k = \frac{Gmm}{R}$$

Where R was the radius of the Sun from its centre to the point where light curved. Newton assumed that deflection angle was actually the ratio of the gravitational potential to the kinetic energy of the particle of light,

$$\theta = \frac{V_r}{E_k}$$

$$\theta = \frac{2GM}{v^2 R}$$

During Newton's time, the speed of the photon (a particle of light) was not known but, today we know this value to a much greater accuracy, thanks to Maxwell and Einstein. All the parameters, from the Sun's mass to the speed of light are known to high accuracy, therefore the Newtonian deflection angle is now known to be,

$$\theta_N = \frac{2GM}{c^2 R} = 0.875 \text{arcsec}$$

The problem with the Newtonian approach is this; we now know that photons of light are massless and move at a constant speed of light c and, the Newtonian deflection angle value is not in agreement with the observations. Therefore Newton's original calculation was flawed and required another explanation.

The Einstein Approach

In the Einstein approach it was found out that, the particles of light were called photons and that these particles where massles moving at a constant speed of light $c = 3 \times 10^8 m/s$. Einstein's theory proposes that gravity is not an actual force, but is instead a geometric distortion of spacetime not predicted by ordinary Newtonian physics. The more mass you have to produce the gravity in a body the more distortion you get, this distortion changes the trajectories of objects moving through space, and even the paths of light rays, as they pass close-by the massive body. Even so, this effect is very feeble for an object as massive as our own sun, so it takes enormous care to even detect that it is occurring.

The Einstein deflection angle was twice the Newton's angle of deflection $\theta = 2\theta_N$, but there is no any account in literature where it shows the derivation of this deflection angle from Einstein field equations, which means that, Einstein came up with a formula similar to the Newtonian deflection angle formula given as,

$$\theta_E = \frac{4GM}{c^2 R} = 1.75 \text{arcsec}$$

Instead of the number, 2 in the Newton formula, we have a 4 and the varying speed of light in the Newton approach is replaced by a constant speed of light c.

The Einstein value was determined by observation through observing the solar eclipse. Although they say it agrees with experiment, we know that this is not true. It has long been suspected that the deflection of light in the vicinity of the sun exceeds the general relativistic predicted value of 1.75". An example of this, is the Erwin Finlay Freundlich 1929 solar eclipse expedition which produced a value of 2.24" larger than the general relativistic value. It is expected that once the reason for the deviation in the deflection angle has been found, it will disprove Einstein's imaginations for the curvature of space time.

It's almost hundred years since Sir Arthur Eddington experimentally proved Einstein's general relativity theory right. Since then, there has never been any competing theory that would prove Einstein wrong save for Loop quantum gravity and string theory. The fact that starlight is bent at the surface of the gravitating body by a deflection angle of 1.75" imposes a bound on the theoretical justification of gravity. Calculating an angle below or above 1.75" will be an upheaval in the founding blocks of physics. Erwin Finlay Freundlich was one of those people who stood out of the ordinary in 1929 when he published results with a larger angle of deflection than Eddington's. An account on Freundlich 1929 expedition has been clearly given in Robert J.Trumpler and Klaus Hentschel papers as stated below;

"Among the various expeditions sent out to observe the total solar eclipse of May 9, 1929, that of the Potsdam Observatory (Einstein Stiftung) seems to be the only one which obtained photographs suitable for determining the light deflection in the Sun's gravitational field. Two instruments were used, but so far only the results of the larger one, a 28-foot horizontal camera combined with a coelostat, have been published. The three observers, Freundlich, von Klüber, and von Brunn, claim that these observations (four plates containing from seventeen to eighteen star images each) lead to a value of 2.24" for the deflection of a light ray grazing the Sun's edge; a figure that deviates considerably from the results of the 1922 eclipse, and which is in contradiction to Einstein's generalized theory of relativity".

The irreducible anomaly in the observations of the deflection of light by the sun has been known to exist since the birth of Einstein General relativity theory. For example, in a 1959 classical review by A.A.Mikhailov, it concludes that observations yield instead of a general relativistic prediction of 1.75arcsec at the limb of the sun the simple mean value of 2.03 ± 0.10 over the GR prediction.

The existence of a 2.24" deflection angle by Freundlich, Von Kluber and Von Brunn therefore implies a requirement for the modification of the general theory of relativity. Science has evolved in this simpler manner of modifications although there are some who cling to the old thoughts of "The earth is the center of the universe and Einstein is always right". I am not proving anyone wrong but I want you to believe that the general relativity theory that was put forward by Einstein is not the only 'there is' excellent description of the universe, there are other ways far better than GR as it was with the Newtonian Gravitational force replacement with a curvature of space time.

The introduction of a number 4 in Einstein deflection angle of light has no basis as to how it came along. The fact that his formula resembles the Newton formula actually shows that Einstein borrowed ideas from Newton analysis. He Einstein also failed to eliminate the mass of a photon from his equations. Even today no one knows how to deduce the deflection angle without taking into account the photon mass because we know the photon is massless.

Ladies and gentlemen, let me present to you another approach that will lead us to the Einstein deflection angle without assuming that the photon has mass or kinetic energy.

Quantum Gravity Approach

Let the potential energy of the Photon according to Einstein –Planck relation be,

$$E = \frac{hc}{\lambda}$$

Where λ is the wavelength

Since ligth appears curved at a small part of the Sun's surface, then the circumference according to deBrogile is quantized in units, $C=\pi R=\lambda$ (In case light orbits the Sun,then $C=2\pi R$). Then the energy of the photon will be given by

$$E_r = \frac{2\hbar c}{R}$$

According to relativistic quantum mechanics, a photon of momentum P, has a kinetic- energy given by, where M is the Sun's mass

$$E_B = \frac{P^2}{2M}$$

According to quantum mechanics in curved space time, space is divided into small chuncks of matter (atoms of space) with a length close to the Planck length l_p , therefore the momentum of a photon passing through these atoms of space will be given by,

$$P = \frac{\hbar}{l_p}$$

This momentum is proof that the photon has no mass and what we percieve as the heaviness of the photon is actually the discrete nature of space.

Due to the discrete nature of space, there is a delay in time at which the photon will reach our telescopes from the distant star. In other words the speed of light doesn't change but there is a huge difference from the calculated time and the observed time of reach of light from the distant star. Then the energy carried by a photon through the discrete space is given as

$$E_B = \frac{\hbar c^3}{2GM}$$

This then brings us to the deflection angle which is the ratio of the photon potential energy to the kinetic energy,

$$\theta = \frac{E_r}{E_B}$$

$$\theta = \frac{4GM}{c^2 R}$$

The Extra Dimension Approach

In higher dimensions or extra dimension problems we get a different picture of what general relativity really is. We assume that light behaves differently in various dimensions and the observations of light from a distant star will vary according to the flux in the extra dimensions because it is this loss of flux to the extra dimensions which makes gravity weak yet it is strong. Therefore what determines our observations is the flux in the extra dimensions as expressed in our model below,

Let the deflection angle of light at the sun's limb be given by ,

$$\theta = \frac{1}{\alpha^{n/2}}\left(\frac{R_s}{R}\right) \qquad (13)$$

Where, $R_s = \frac{2GM}{c^2}$ is the Schwarzschild radius of a gravitating body, α is the size of the extra dimension and $\alpha^{n/2}$ is the flux in the extra dimension. In what follows, we use the above equation by subsituting in the values of $\alpha^{n/2}$ to get the values of the three deflection angles whose sample mean gives the Einstein deflection value. This analysis will help us recover new theories based on the flux in the extra dimension.

Let us start with the Newton's theory of gravitation. To recover the Newtonian deflection angle at the suns limb, we set $\alpha^{n/2} = 1$. This then gives the Newtonian value as,

$$\theta_N = \frac{R_s}{R_\odot} = 0.875 arcsec$$

The Freundlich deflection angle might have taken a different twist than with Eddington 1.75arcsec result, which we are yet to find out. Taking, $\alpha^{n/2} = 0.0233$, we deduce the deflection angle given by,

$$\theta_F = \frac{2.56R_s}{R_\odot} = 2.24 \text{arcsec}$$

Lastly when $\alpha^{n/2} = 0.0290$ we get the following deflection angle,

$$\theta_Q = \frac{2.426R_s}{R_\odot} = 2.12 \text{arcsec}$$

Our first result from the above calculations is that; the sample mean of the deflection angles from the three observations gives the exact deflection angle that was calculated and observed by Eddington in General relativity as,

$$\frac{\sum_{n=1}^{4} \theta_n}{3} = \frac{0.875 + 2.24 + 2.12}{3}$$
$$= 1.75 \text{arcsec}$$

The fact that the mean of the three observations for the deflection of light given above reproduces the GR value of 1.75arcsec sums up what exactly general relativity really is. In simple terms GR is the sample mean of three observations taken from different location on the earth's surface where the flux in the extra dimension makes the strength of gravity slightly different in those positions where light bends.

The model given above is proof that the curvature of space assumption given by General Relatitiy was just a mathematical artifact and not a real entity. The observed deflection angles are greatly determined by the flux in the extra dimensions.

A new Approach to Quantum Theory

The Wave and Particle Analogy

Basing our study on the electric currents generated whenever there is a changing magnetic field (B) and a changing electric field (E) in the electromagnetic wave we can construct a complete theory for the electromagnetic radiations. The theory is created using the symmetry between a long wire placed in the electromagnetic fields which induce vibrating electrons that carry current in the wire and the electromagnetic wave which constitute changing electric and magnetic fields that create vibrating photons in the wave.

Therefore a wire is to a wave what a vibrating electron is to a vibrating photon in the wire and a wave respectively. The aim of the paper is to give a clear description of the theory of electromagnetic radiations (light).

The goal of the book on the other hand is to show that the wave-particle descriptions of reality can be applied to any physical situation simultaneously.

The objective of the book is to show that the Photoelectric Effect and the Compton Effect can both be explained by the wave model and the particle model at the same time.

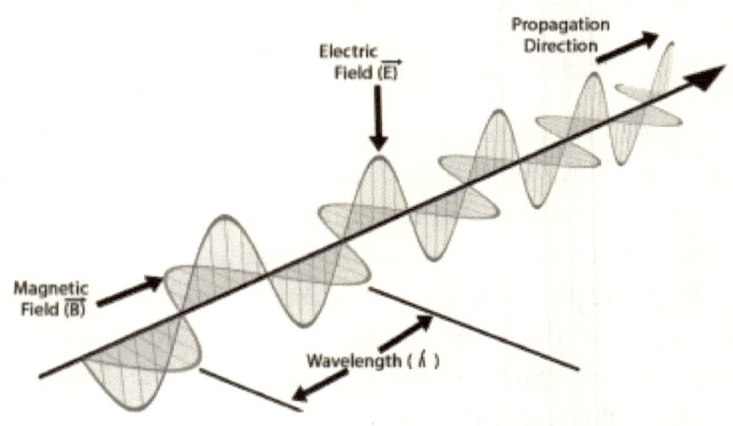

Consider a long wire connected to an ammeter and strong electric and magnetic fields produced in a vacuum. Let us assume that whenever a wire is brought in vicinity of a changing electric field, electrons of mass (m) are set into motion in the wire and then an ammeter deflects, recording a current (i_E). The current in the wire due to a changing electric field should be given by

$$i_E = \frac{j\varepsilon_0}{2\pi m}$$
$$E\,(1)$$

Where (ε_0) is the permittivity of free space and (j) is the constant of action in SI units Js. therefore the current is quantized and depends on both the electric field and the mass of an electron.

When the wire is brought into the magnetic field, vibrating electrons at a frequency of oscillation (f) are set in motion at a speed (v) through the wire generating a current given by

$$i_B = \frac{v}{2\pi \mu_o f} B$$

$$(2)$$

Where (μ_o) is the permeability of free space.

Assuming that the ammeter records different values of (i_E) and (i_B), what will be the change in the current values recorded at the ammeter? Subtracting equation (1) from equation (2) we have

$$\Delta I = (i_E - i_B) = \left(\frac{J\varepsilon_o}{2\pi m} E - \frac{v}{2\pi \mu_o f} B \right)$$

$$(3)$$

This is the change in the currents due to changing magnetic and electric fields. Assuming that there is no change in the current, meaning that the current values for i_E are equal to those of i_B (i.e $\Delta I = 0$). This will imply that the magnetic field strength was equal to the electric field strength at one point in both experiments. In terms of electromagnetic radiations in the vacuum, assuming that a wire carrying current is replaced by a wave and electrons are replaced by photons. The wire replaced by a wave is made up of vibrating electric and magnetic fields at a given frequency making an electromagnetic wave. The electrons replaced with photons will represent the particle properties of the electromagnetic wave (light) with associated mass and speed (v).

The symmetry here is between the long wire and the wave, the electrons and the Photons. The electric and magnetic fields brought in vicinity of the wire and the number of oscillations per second of the electron in the wire is what leads to an electromagnetic wave. The electrons with a given mass and moving at a given speed is what constitute a photon. Then at $\Delta I = 0$, we have on arranging,

$$\frac{jf}{mv} = \frac{1}{2\pi\mu_0\varepsilon_0}\frac{B}{E}$$

$$(4)$$

This means that at $\Delta l = 0$, either a changing magnetic field or a changing electric field produces a current. Then it should be true that a changing magnetic field produces an electric field just as a changing electric field produces a magnetic field. This process in the electromagnetic wave continues indefinitely. The electromagnetic wave will move at a constant speed (c), since for electromagnetic waves, $\frac{E}{B} = c$, and for a photon $\frac{jf}{mv} = c$ where j=6.63×10^{-34} Js (also called the Planck constant after Max Planck) and mv is the photon momentum. Implying that the photon energy is related to the frequency of the electromagnetic wave by (jf). Then the electromagnetic wave will move at a constant speed given as, since by symmetry $\frac{E}{B} = \frac{jf}{mv} = c$

$$c = \frac{1}{\sqrt{\varepsilon_0\mu_0}} = 2.99792458 \times 10^8 \frac{m}{s}$$

Where $\varepsilon_o = 8.85418782 \times 10^{-12}\frac{c^2}{Nm^2}$ and $\mu_o = 1.26 \times 10^{-6}\frac{Ns^2}{c^2}$

We have therefore deduced based on the symmetry between a current (electron) carrying wire in the electromagnetic field and the photons in electromagnetic waves that an electromagnetic wave moves at a constant speed of light. It is also true from the deductions that light is indeed made up of particles of light called photons and vibrating electric and magnetic fields. The deduction would not be possible if the wave and particle descriptions of the situations had not been applied simultaneously (into what is called "the wave-particle duality).

The Photoelectric Effect

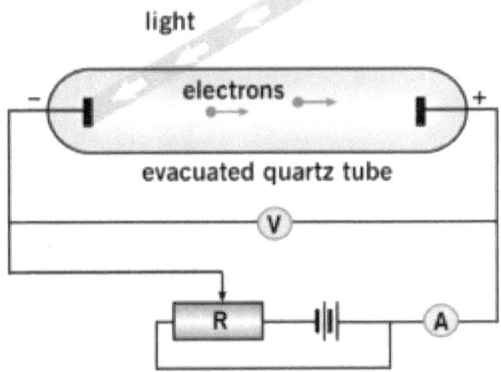

Unexpectedly enough the photoelectric effect can also be explained by Equation (3), on arranging

$$\frac{2\pi m f}{\varepsilon_0 E} \Delta l = jf - \frac{mv}{2\pi\,\mu_0\varepsilon_0}\frac{B}{E}$$

Then the total energy of the particle of light (Photon) is then given by

$$jf = \frac{2\pi m f}{\varepsilon_0 E} \Delta l + \frac{mv}{2\pi\,\mu_0\varepsilon_0}\frac{B}{E}$$

$$(5)$$

It is therefore true that the photoelectric effect can be explained when both the particle and wave models of reality are applied in the experiment at the same time (simultaneously). The work function from Einstein's photoelectric equation (A. Einstein, 1905) will here be replaced by $\frac{2\pi m f}{\varepsilon_0 E} \Delta l$ while the kinetic energy of the electrons at the surface of the metal will be given by $\frac{mv}{2\pi\mu_0\varepsilon_0}\frac{B}{E}$. Equation (5) reduces to Einstein's Photoelectric effect when, the speed of the electron is $v = \frac{1}{\pi \mu_0 \varepsilon_0} \frac{B}{E}$ and the change in current for a complete circuit is $\Delta I = \frac{j \varepsilon_0 E}{2 \pi m}$.

The Compton Effect

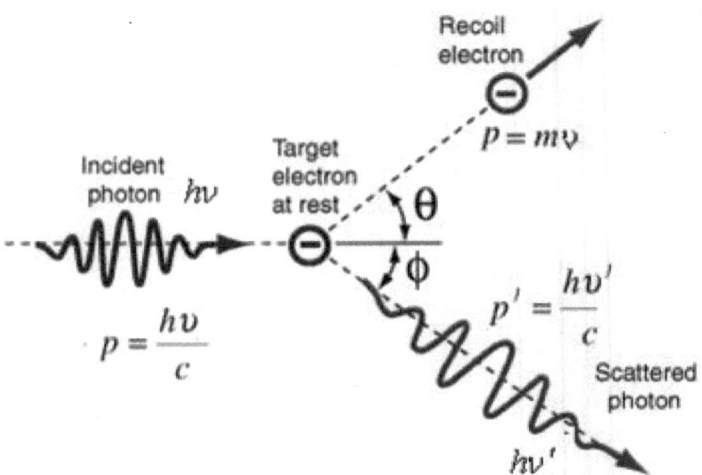

The validity of the Compton Effect can also be deduced from Equation (3). The current can be taken as the product of the frequency (f) of radiations and the charge (q) on the particle.

Then the current due to the electric field is $i_E = qf_1$ and that due to the magnetic field is $i_B = qf_2$. In the case of the Compton Effect, q is the charge on the free electron while f_1 and f_2 are the frequencies of the incoming photon and outgoing photon after collision with the free electron respectively. Then equation (3) can be written as

$$f_1 - f_2 = \frac{1}{q}\left(\frac{j\varepsilon_0}{2\pi m}E - \frac{v}{2\pi\mu_0 f}B\right)$$
(6)

Since photons move with the speed of light(c) then their frequencies is related to their speed and wavelength by $f = \frac{c}{\lambda}$, then we have

$$\frac{1}{\lambda_1} - \frac{1}{\lambda_2} = \frac{1}{qc}\left(\frac{j\varepsilon_0}{2\pi m}E - \frac{v}{2\pi\mu_0 f}B\right)$$

On arranging to include the charge density of the free electron for electric field lines in an area of $\frac{\lambda_1\lambda_2}{2\pi}$, we obtain

$$\frac{2\pi q}{\lambda_1\lambda_2}(\lambda_2 - \lambda_1) = \frac{j}{mc}\left(\varepsilon_0 E - \frac{mv}{\mu_0 jf}B\right)$$

Where (mc) is the momentum of an electron treated relativistic ally, on letting the charge density $= \frac{2\pi q}{\lambda_1 \lambda_2} = \varepsilon_0 E$, we deduce the change in the wave length of the incoming photon and outgoing photon after collision with the free electron as

$$\Delta\lambda = (\lambda_2 - \lambda_1) = \frac{j}{mc}\left(1 - \frac{mv}{\rho\mu_0 jf}B\right)$$

Since $\rho = \varepsilon_0 E$, we then have

$$\Delta\lambda = (\lambda_2 - \lambda_1) = \frac{j}{mc}\left(1 - \frac{\frac{mvB}{\varepsilon_0\mu_0 E}}{jf}\right)$$

Since jf is the energy carried by the photon, and then also $\frac{mvB}{\varepsilon_0\mu_0 E}$ is the energy carried by the free electron. Treating the electron relativistically such that for electromagnetic waves moving at a speed (v) relative to the electron moving at a speed of light $c = \frac{1}{\sqrt{\varepsilon_0\mu_0}}$, the electric field in the wave will be related to the magnetic field by $Bv = E.$ then the energy carried by an electron can be given by mc^2. Then the angle at which the photon is scattered after collision with the free electron will be given by

$$\theta = \cos^{-1}\left(\frac{mvB}{\varepsilon_0\mu_0 E}\right)\Big/ jf$$

(7)

Where mv is the momentum of the photon in the electromagnetic wave consisting of a changing electric field E and magnetic field B both moving at a constant speed of light $c = \dfrac{1}{\sqrt{\varepsilon_0 \mu_0}}$. Treating the electron relativistically we have

$$\theta = \cos^{-1} \frac{mc^2}{jf}$$

When the energy carried by the photon is equal to the energy possessed by the electron then $\theta = 0$, meaning that there is or there is no scattering and whatsoever there is no increase in photon wavelength hence $\Delta\lambda = 0$.

Bohr's Atomic Theory

A complete theory of light can't fail to explain the structure of an atom. I therefore take a complete discussion of what goes on inside an atom only with the help of Bohr's energy levels which he derived using classical mechanics and quantum theory. Let $\Delta f = f_1 - f_2$ be an increase in the frequency of the electromagnetic radiations emitted from an atom. Then squaring both sides of equation (6) and arranging will give

$$4\pi^2 \Delta f^2 = \frac{1}{m^2 q^2}\left(j\varepsilon_o E - \frac{mv}{\mu_o f}B\right)^2$$

$$4\pi^2 m^2 q^2 \Delta f^2 = j^2 \varepsilon_o^2 E^2 - 2\frac{j\varepsilon_o EBmv}{\mu_o f} + \frac{B^2 m^2 v^2}{\mu_o^2 f^2}$$

Dividing through by $64\pi^4 j^2 \varepsilon_o^2$ and multiplying through by q^2 gives the energy of the atom as on arranging

$$\frac{mq^4}{16\pi^2 \pi^4 j^2 \varepsilon_o^2} = \frac{1}{64\pi^4 m \Delta f^2}\left((Eq)^2 - 2\frac{m(Eq)(Bqv)}{\mu_o \varepsilon_o(jf)} + \frac{(Bqv)^2 m^2}{\mu_o^2 \varepsilon_o^2 (jf)^2}\right)$$

The energy of the n-th level is since the reduced Planck constant is

$$n\hbar = \frac{nj}{2\pi}$$

$$\frac{mq^4}{32\pi^2 \pi^4 n^2 \hbar^2 \varepsilon_o^2} = \frac{1}{32\pi^2 n^2 m \Delta f^2}\left((Eq)^2 - 2\frac{m(Eq)(Bqv)}{\mu_o \varepsilon_o(jf)} + \frac{(Bqv)^2 m^2}{\mu_o^2 \varepsilon_o^2 (jf)^2}\right)$$

The expression on the left hand side of the equation is the quantized energy of an atom (Niels Bohr, 1913) while the right hand side of the equation represents the energy of the atom in terms of the forces associated with it. In the equation we let $H_e = Eq$ be the electric force for a particle moving in the electric field and $H_b = Bqv$, the magnetic force on a particle with charge q moving in the magnetic field. Since the speed of light is $= \frac{1}{\sqrt{\varepsilon_0 \mu_0}}$, then the quantized energy can be given as

$$W_n = \frac{1}{32\pi^2 n^2 m \Delta f^2} \left(H_e^2 - 2\frac{H_e H_b mc^2}{jf} + \frac{H_b^2 (mc^2)^2}{(jf)^2} \right)$$

Then on arranging we obtain

$$W_n = \frac{1}{32\pi^2 n^2 m \Delta f^2} \left(H_e - \frac{mc^2}{jf} H_b \right)^2$$

$$(8)$$

When the energy of an electron moving at a speed of light in atom is equal to the energy of the emitted photon, then

$$W_n = \frac{1}{32\pi^2 n^2 m \Delta f^2} (H_e - H_b)^2 = \frac{1}{32\pi^2 mn^2} \left(\frac{\Delta H}{\Delta f} \right)^2 \quad (9)$$

Where $\Delta H = H_e - H_b$ is the difference or change between the electric force and the magnetic force in an atom, when the two forces balance (i.e. $H_e = H_b$), then $W_n = 0$ meaning that the total energy of an atom will cease to exist.

Therefore the total energy of an atom increases with the square of the change in the electric and magnetic forces which govern an electron but falls off as the square of the change in the frequency of the radiation emitted by it.

From equation (8) the ratio of the energy of an electron to that
of the photon $\frac{mc^2}{jf}$, is the limit at which if the energies are not
equal you will not get a change in the electric and magnetic forces.
Treating the ratio as a number $\tau = \frac{mc^2}{jf}$, we get from equation (8)

$$W_n = \frac{1}{32\pi^2 mn^2}\left(\frac{H_e - \tau H_b}{f_1 - f_2}\right)^2$$

(10)

When $\tau = 0$, it means that the relativistic energy (mc^2) of an
electron in an atom is zero, and that the total energy of an atom
only increases with the electric force on the electron. The
relationship (equation 10) is a complete expression for the laws
according to which, by the theory here advanced, the structure of
an atom should be viewed.

In conclusion, a complete theory of light is only possible if both the
wave and particle descriptions of reality are applied to the physical
situation at the same time. In discussing Young's double slit
experiment for example we should be able with the formulas given
above to treat the electromagnetic radiations on both a wave and
particle model.

The Volume Entropy Law of a Black Hole

The development of general relativity followed a publication of acceleration under special relativity in 1907 by Albert Einstein. In his article, he argued that any mass will "Distort" the region of space around it so that all freely moving objects will follow the same curved paths curving toward the mass producing the distortions. Then in 1916, Schwarzschild found a solution to the Einstein field equations, laying the groundwork for the description of gravitational collapse and eventually black holes.

A black hole is created when a dying star collapses to a singular point, concealed by an "event horizon;" the black hole is so dense and has such strong gravity that nothing, including light, can escape it; black holes are predicted by general relativity, and though they cannot be "seen," several have been inferred from astronomical observations of binary stars and massive collapsed stars at the centers of galaxies.

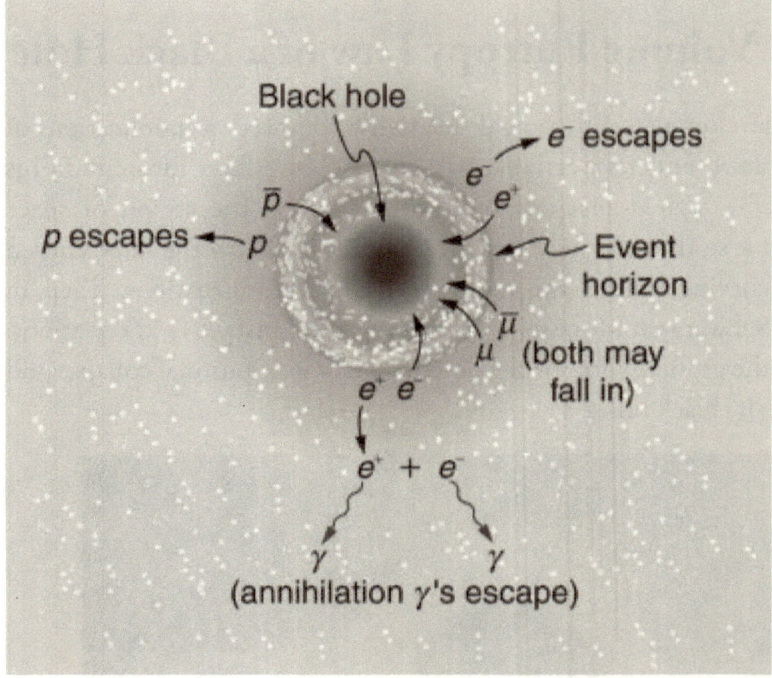

These objects have puzzled the minds of great thinkers for many years. History puts it that, they were first predicated by John Michell and Pierre-Simon Laplace in the 18th century but David Finkelstein was the first person to publish a promising explanation of them in 1958 based on Karl Schwarz child's formulations of a solution to general relativity that characterized black holes in 1916.

In 1971, Hawking developed what is known as the second law of black hole mechanics in which the total area of the event horizons of any collection of classical black holes can never decrease, even if they collide and merge. This was similar to the second law of thermodynamics which states that, the entropy of a system can never decrease.

Then in 1972 Bekenstein proposed an analogy between black hole physics and thermodynamics in which he derived a relation between the entropy of black hole entropy and the area of its event horizon.

$$S = \frac{Akc^3}{4G\hbar}$$

In 1974, Hawking predicted an entirely astonishing phenomenon about black holes, in which he ascertained with accuracy that black holes do radiate or emit particles in a perfect black body spectrum.

$$T = \frac{\hbar c^3}{8\pi GMk}$$

The Bekenstein-Hawking area entropy law raises a number of questions. Why does the entropy of a Black hole scale with its area and not with its volume? For systems that we have studied, the entropy is proportional to the volume of the system. If entropy is proportional to area, so what do we make of all those thermodynamic relations that include volume, like Boyle's law or descriptions for a gas in a box? In otherwords how do we associate volume to the entropy of a Black hole?

Area Entropy Law

A Black hole is defined as a mathematical spacetime singularity that is; a position in space where quantities used to determine the gravitational field become infinite; such quantities include the curvature of spacetime and the density of matter. That is, for high or infinite densities where matter is enclosed in a very small volume of space General relativity breaks down.

Quantum mechanics suggests that there may be no such thing in nature as a point in space-time, implying that space-time is always smeared out, occupying some minimum region. The minimum smeared-out volume of space-time is a profound property in any quantized theory of gravity and such an outcome lies in a widespread expectation that singularities will be resolved in a quantum theory of gravity. Therefore associating area to the entropy of a black hole means that the black hole has no volume which may not be true in a theory of quantum mechanics in curved space-time.

If density is the amount of energy contained within a given volume of space, then a Black hole must have a density and its volume will be determined by the amount of space enclosed by the surface area of its event horizon. Then from our energy density equation (see previous series) we can derive the Area entropy law of a Black hole. Let the intensity I of the radiation emitted by a black hole when the amount of energy is added be

$$I = \frac{F^2}{8\pi\hbar}$$

Intensity is power per unit surface area of a Black hole A and power is the amount of energy E added to a black hole in time t. This implies that the energy or heat added to a black hole is

$$E = \frac{AtF^2}{2\pi\hbar}$$

Where $\hbar$ is the reduced Planck constant

From Einstein General relativity, the gravitational force felt by a body near the black hole is very strong and is given by,

$$F = \frac{c^4}{4G}$$

From the time-energy Uncertainty Principle, the time taken by the black hole of mass M to dissipate is,

$$t = \frac{\hbar}{Mc^2}$$

Then on substitution into the energy formula we have

$$W = \frac{Ac^6}{32\pi G^2 M}$$

This is the Frodden-Ghosh- Perez Energy.

Hawking knew that if the horizon area were an actual entropy, black holes would have to radiate. When heat is added to a thermal system, the change in entropy is the increase in thermal energy divided by temperature:

$$S = \frac{E}{T}$$

From Hawking original calculation, the temperature of a black hole is given by,

$$T = \frac{\hbar c^3}{8\pi GMk}$$

Therefore the Entropy of a Black hole will be given by,

$$S = \frac{\left(\dfrac{Ac^6}{32\pi G^2 M}\right)}{\left(\dfrac{\hbar c^3}{8\pi GMk}\right)} = \frac{Akc^3}{4G\hbar}$$

This implies that the heat added into the black hole goes into increasing its area and an increase in area will automatically lead to an increase in entropy as explained by the above given area entropy law.

Black hole Volume Entropy law

If density is the amount of energy contained within a given volume of space, then a Black hole must have a density and its volume will be determined by the amount of space enclosed by the surface area of its event horizon. Let the energy density of a black hole be given as the work done on the system by exterior agents (E) per unit volume (V),

$$\rho = \frac{E}{V} = \frac{F^2}{2\pi\hbar c}$$

(Note: Instead of the Intensity formula we have used the energy density formula)

As before, the strong gravitational force of a black hole is

$$F = \frac{c^4}{4G}$$

The energy added to the black hole is related to the Volume of the black hole by;

$$E = \frac{Vc^7}{32\pi\hbar G^2}$$

This energy is very different from the Frodden-Ghosh- Perez Energy. This means that the energy added goes into increasing the volume of the Black hole not its area.

Hawking knew that if the horizon area were an actual entropy, black holes would have to radiate. When heat is added to a thermal system, the change in entropy is the increase in thermal energy divided by temperature:

$$S = \frac{E}{T}$$

From Hawking original calculation, the temperature of a black hole is given by,

$$T = \frac{\hbar c^3}{8\pi GMk}$$

Therefore the Entropy of a Black hole will be given by,

$$S = \frac{\left(\dfrac{Vc^7}{32\pi\hbar G^2}\right)}{\left(\dfrac{\hbar c^3}{8\pi GMk}\right)} = \frac{VkMc^4}{4\hbar^2 G}$$

Therefore the entropy of a black hole is related to the Volume and mass of a black hole by the above given formula. Therefore an increase in both mass-energy Mc^2 and volume of the black hole leads to an increase in entropy as

$$S = V(Mc^2)\frac{kc^2}{4\hbar^2 G}$$

However in a limit where the Volume of the black hole is related to its area A, that is for Black holes the size of Debrogile wavelength

$$\lambda = \frac{\hbar}{Mc}$$

$$V = \frac{\hbar A}{Mc}$$

We recover the Bekenstein Hawking area entropy law.

Finally when the black hole has a mass equal to the Planck mass particle

$$M = \sqrt{\frac{\hbar c}{G}}$$

We get the entropy of the black hole to be

$$S = \frac{Vk}{4}\left(\frac{c^3}{G\hbar}\right)^{3/2}$$

Space-time Singularity Resolution

It has been known for some time that a star more than three times the size of our Sun collapses in this way, the gravitational forces of the entire mass of a star overcomes the electromagnetic forces of individual atoms and so collapse inwards. If a star is massive enough it will continue to collapse creating a Black hole, where the whopping of space time is so great that nothing can escape not even light, it gets smaller and smaller. The star in fact gets denser as atoms even subatomic particles literally get crashed into smaller and smaller space, and its ending point is of course a space time singularity.

In summary, a Black hole is that object created when a dying star collapses to a singular point, concealed by an event horizon, it is so dense and has strong gravity that nothing, including light, can escape it. Black holes are predicted by general relativity, and though they cannot be "seen," several have been inferred from astronomical observations of binary stars and massive collapsed stars at the centers of galaxies.

Black holes formed by gravitational collapse require great energy density but there exists a new breed of Black holes that where formed in the early universe after the big bang, where the energy density was much greater allowing the formation of Primordial Black holes with masses ranging from, 10^8 , 10^{12} - 10^{17}kg . Therefore the formation of primordial, min or quantum black holes was due to density perturbations forming in it a gravitational collapse in the early universe.

A Black hole might not actually be a physical object in space but rather a mathematical singularity, a prediction of Einstein's General Relativity theory, a place where the solutions of Einstein differential equations break down. A space-time singularity therefore is a position in space where quantities used to determine the gravitational field become infinite; such quantities include the curvature of space-time and the density of matter. Singularities are places where both the curvature and the energy-density of matter become infinitely large such that light cannot escape them. This happens for example inside black holes and at the beginning of the early universe.

Singularities in any physical theory indicate that either something is wrong or we need to reformulate the theory itself. Singularities are like dividing something by zero.

The problems that plague the General relativity theory arise from trying to deal with a point in space or a universe that is zero in size (infinite densities). However, quantum mechanics suggests that there may be no such thing in nature as a point in space-time, implying that space-time is always smeared out, occupying some minimum region. The minimum smeared-out volume of space-time is a profound property in any quantized theory of gravity and such an outcome lies in a widespread expectation that singularities will be resolved in a quantum theory of gravity. This implies that the study of singularities acts as a testing ground for quantum gravity.

Loop quantum gravity (LQG) suggests that singularities may not exist. LQG states that due to quantum gravity effects, there must be a minimum distance beyond which the force of gravity no longer continues to increase as the distance between the masses become shorter or alternatively that interpenetrating particle waves mask gravitational effects that would be felt at a distance. It must also be true that under the assumption of a corrected dynamical equation of LQ cosmology and brane world model, for the gravitational collapse of a perfect fluid sphere in the commoving frame, the sphere does not collapse to a singularity but instead pulsates between a maximum and minimum size, avoiding the singularity.

Additionally, the information loss paradox is also a hot topic of theoretical modeling right now because it suggests that either our theory of quantum physics or our model of black holes is flawed or at least incomplete. and perhaps most importantly, it is also recognized with some prescience that resolving the information paradox will hold the key to a holistic description of quantum gravity, and therefore be a major advance towards a unified field theory of physics.

Singularities are a sign that the theory breaks down and has to be replaced by a more fundamental theory. And we think the same has to be the case in General Relativity, where the more fundamental theory to replace it is quantum gravity.

Whether in gravitational collapse or the early universe, we now know that the formation of Black holes or space time singularities requires great and much greater energy density. This we know because while the left hand side of Einstein field equations representsnts the metric of space-time curvature, the right hand side represents the matter- energy content of the classical matter fields of pressure and energy density. This therefore means that quantum mechanics which plays an important role in the behavior of the matter fields has no place in the Einstein field equations and this is what brings on the singularities that plague the general relativity theory.

$$G_{\mu\nu} + \Lambda g_{\mu\nu} = \frac{8\pi G}{c^4} T_{\mu\nu}$$

Because of this, one therefore has a problem of defining a consistent scheme in which the space time metric is treated classically but is coupled to the matter fields which are treated quantum mechanically.

The approximation I shall use on my journey to quantum gravity (Quantum Black holes) is that the matter fields, such as scalar, electro-magnetic, or neutrino fields, obey the usual wave equations with the left hand side replaced by a classical space time second order curvature

$$\Lambda = \frac{1}{R^2}$$

Where R is the radius of curvature, while the right hand stress-energy tensor is replaced by the quantum mechanical energy density (see Previous ThinkPhysics serious)

$$\rho = \frac{F^2}{8\pi\alpha\hbar c} \qquad (1)$$

Where F is the force involved in an interaction α is the coupling constant that determines the strength of the force, and $\hbar$ is the reduced Planck constant. The equation represents the coupling constant (α) as a function of the energy density (ρ) for any force (F) exerted in an interaction. The application of this equation is the Franzl Aus Tirol curve on Wikipedia's "Coupling constant". Another application is the derivation of energy stored in the electromagnetic field. Therefore the general theory of quantum mechanics in curved space –time will be given by this simple equation,

$$\Lambda = \frac{8\pi G}{c^4}\rho$$

Where,

$$\Lambda = \frac{GF^2}{\alpha\hbar c^5} = \frac{F^2}{\alpha E_{pl}^2} \qquad (2)$$

Where, $E_{pl} = M_{pl}c^2$ is the Planck energy and M_{pl} is the Planck mass

From the above given equation we see that high space curvature will always be achieved when the square of the force involved increases. According to the theory given, this will only occur at the Planck energy level where space is discrete or granular in nature (its building blocks being exactly the Planck mass, simply put, the atoms of space). There is no change in energy because the only energy involved in the process is the constant Planck energy of the Planck mass.

As we said earlier, that the formation of a black hole due to the process of gravitational collapse occurs in the presence of great energy density and also that the formation of primordial black holes in the early universe occurs in the presence of a much greater energy density, our theory suggests that this energy density is high because of the strong gravitational force involved in the process. According to general relativity, this force is a constant and is given by

$$F = \frac{c^4}{G}$$

Therefore from equation (2), when this force is present the curvature of space scales as the inverse of the square of the Planck length,

$$\Lambda = \frac{c^3}{a\hbar G} = \frac{1}{a l_p^2} \qquad (3)$$

Where $l_p = \sqrt{\frac{\hbar G}{c^3}}$ is the Planck length.

This implies that, in the theory of quantum mechanics in curved space-time for the gravitational collapse of a star, the star does not collapse to a singularity but instead to a Planck sized star of Planck length close to 10^{-35}m and this will happen only when $\alpha = 1$ .Finally, in the theory of quantum mechanics in curved space-time, we consider the possibility that the energy of a collapsing star and any additional energy falling into the hole could condense into a highly compressed core with density of the order of the Planck density. Since the energy density or pressure is expressed as in equation (1),

$$\rho = \frac{F^2}{8\pi a \hbar c}$$

Therefore nature appears to enter the quantum gravity regime when the energy density of matter reaches the Planck scale. The point is that this may happen well before relevant lengths become planckian. For instance, a collapsing spatially compact universe bounces back into an expanding one. The bounce is due to a quantum-gravitational repulsion which originates from the modified Heisenberg uncertainty, and is akin to the force that keeps an electron from falling into the nucleus. And from the uncertainity principle, this repulsion force is given by,

$$F = \frac{c^4}{G}$$

Therefore the bounce does not happen when the universe is of planckian size, as before; it happens when the matter energy density reaches the Planck density in this way,

$$\rho = \frac{c^7}{8\pi\alpha\hbar G^2} \qquad (4)$$

At this energy density, a Planck star is formed. The key feature of this theoretical object is that this repulsion arises from the energy density, not the Planck length, and starts taking effect far earlier than might be expected. This repulsive 'force' is strong enough to stop the collapse of the star well before a singularity is formed, and indeed, well before the Planck scale for distance. Since a Planck star is calculated to be considerably larger than the Planck scale for distance, this means there is adequate room for all the information captured inside of a black hole to be encoded in the star, thus avoiding information loss.

The analogy between quantum gravitational effects on

Cosmological and black-hole singularities has been exploited to study if and how quantum gravity could also resolve the $r = 0$ singularity at the center of a collapsed star, and there are good indications that it does. For example, if we extend (3) to n extra dimensions we have,

$$R = \alpha^{n/2} l_p$$

Where α in this case is the size of the extra dimensions and α^n is the flux in the extra dimesions. Let the size of the extra dimension be given as the gravitational coupling constant, $\alpha = \frac{GM^2}{\hbar c} = \left(\frac{M}{M_{pl}}\right)^2$, then the size of a star will be given by,

$$r = \left(\frac{M}{M_{pl}}\right)^n l_p \qquad (5)$$

Where M is the mass of the star and n is positive. For instance, if n = 1/3, a stellar-mass black hole would collapse to a Planck star with a size of the order of 10^{-10} centimeters. This is very small compared to the original star in fact, smaller than the atomic scale but it is still more than 30 orders of magnitude larger than the Planck length. This is the scale on which we are focusing here. The main hypothesis here is that a star so compressed would not satisfy the classical Einstein equations anymore, even if huge compared to the Planck scale because its energy density is already planckian.

Glossary

Absolute space and time—the Newtonian concepts of space and time, in which space is independent of the material bodies within it, and time flows at the same rate throughout the universe without regard to the locations of different observers and their experience of "now."

Acceleration—the rate at which the speed or velocity of a body changes.

Accelerating universe—the discovery in 1998, through data from very distant supernovae, that the expansion of the universe in the wake of the big bang is not slowing down, but is actually speeding up at this point in its history; groups of astronomers in California and Australia independently discovered that the light from the supernovae appears dimmer than would be expected if the universe were slowing down.

Action—the mathematical expression used to describe a physical system by requiring only the knowledge of the initial and final states of the system; the values of the physical variables at all intermediate states are determined by minimizing the action.

Anthropic principle—the idea that our existence in the universe imposes constraints on its properties; an extreme version claims that we owe our existence to this principle.

Asymptotic freedom (or safety)—a property of quantum field theory in which the strength of the coupling between elementary particles vanishes with increasing energy and/or decreasing distance, such that the elementary particles approach free particles with no external forces acting on them; moreover for decreasing energy and/or increasing distance between the particles, the strength of the particle force increases indefinitely.

Baryon—a subatomic particle composed of three quarks, such as the proton and neutron.

Big bang theory—the theory that the universe began with a violent explosion of spacetime, and that matter and energy originated from an infinitely small and dense point.

Big crunch—similar to the big bang, this idea postulates an end to the universe in a singularity.

Binary stars—a common astrophysical system in which two stars rotate around each other; also called a "double star."

Blackbody—a physical system that absorbs all radiation that hits it, and emits characteristic radiation energy depending upon temperature; the concept of blackbodies is useful, among other things, in learning the temperature of stars.

Black hole—created when a dying star collapses to a singular point, concealed by an "event horizon;" the black hole is so dense and has such strong gravity that nothing, including light, can escape it; black holes are predicted by general relativity, and though they cannot be "seen," several have been inferred from astronomical observations of binary stars and massive collapsed stars at the centers of galaxies.

Boson—a particle with integer spin, such as photons, mesons, and gravitons, which carries the forces between fermions.

Brane—shortened from "membrane," a higher-dimensional extension of a onedimensional string.

Cassini spacecraft—NASA mission to Saturn, launched in 1997, that in addition to making detailed studies of Saturn and its moons, determined a bound on the variations of Newton's gravitational constant with time.

Causality—the concept that every event has in its past events that caused it, but no event can play a role in causing events in its past.

Classical theory—a physical theory, such as Newton's gravity theory or Einstein's general relativity, that is concerned with the macroscopic universe, as opposed to theories concerning events at the submicroscopic level such as quantum mechanics and the standard model of particle physics.

Copernican revolution—the paradigm shift begun by Nicolaus Copernicus in the early sixteenth century, when he identified the sun, rather than the Earth, as the center of the known universe.

Cosmic microwave background (CMB)—the first significant evidence for the big bang theory; initially found in 1964 and studied further by NASA teams in 1989 and the early 2000s, the CMB is a smooth signature of microwaves everywhere in the sky, representing the "afterglow"of the big bang: Infrared light produced about 400,000 years after the big bang had redshifted through the stretching of spacetime during fourteen billion years of expansion to the microwave part of the electromagnetic spectrum, revealing a great deal of information about the early universe.

Cosmological constant—a mathematical term that Einstein inserted into his gravity field equations in 1917 to keep the universe static and eternal; although he later regretted this and called it his "biggest blunder," cosmologists today still use the

cosmological constant, and some equate it with the mysterious dark energy.

Coupling constant—a term that indicates the strength of an interaction between particles or fields; electric charge and Newton's gravitational constant are coupling constants.

Crystalline spheres—concentric transparent spheres in ancient Greek cosmology that held the moon, sun, planets, and stars in place and made them revolve around the Earth; they were part of the western conception of the universe until the Renaissance.

Curvature—the deviation from a Euclidean spacetime due to the warping of the geometry by massive bodies.

Dark energy—a mysterious form of energy that has been associated with negative pressure vacuum energy and Einstein's cosmological constant; it is hypothesized to explain the data on the accelerating expansion of the universe; according to the standard model, the dark energy, which is spread uniformly

throughout the universe, makes up about 70 percent of the total mass and energy content of the universe.

Dark matter—invisible, not-yet-detected, unknown particles of matter, representing about 30 percent of the total mass of matter according to the standard model; its presence is necessary if Newton's and Einstein's gravity theories are to fit data from galaxies, clusters of galaxies, and cosmology; together, dark

matter and dark energy mean that 96 percent of the matter and energy in the universe is invisible.

Deferent—in the ancient Ptolemaic concept of the universe, a large circle representing the orbit of a planet around the Earth.

Doppler principle or **Doppler effect**—the discovery by the nineteenth-century Austrian scientist Christian Doppler that when sound or light waves are moving toward an observer, the apparent frequency of the waves will be shortened, while if they are moving away from an observer, they will be lengthened; in astronomy this means that the light emitted by galaxies moving away from us is redshifted, and that from nearby galaxies moving toward us is blueshifted.

Dwarf galaxy—a small galaxy (containing several billion stars) orbiting a larger galaxy; the Milky Way has over a dozen dwarf galaxies as companions, including the Large Magellanic Cloud and Small Magellanic Cloud.

Dynamics—the physics of matter in motion.

Electromagnetism—the unified force of electricity and magnetism, discovered to be the same phenomenon by Michael Faraday and James Clerk Maxwell in the nineteenth century.

Electromagnetic radiation—a term for wave motion of electromagnetic fields which propagate with the speed of light—300,000 kilometers per second—and differ only in wavelength; this includes visible light, ultraviolet light, infrared radiation,

X-rays, gamma rays, and radio waves.

Electron—an elementary particle carrying negative charge that orbits the nucleus of an atom.

Eötvös experiments—torsion balance experiments performed by Hungarian Count Roland von Eötvös in the late nineteenth and early twentieth centuries that showed that inertial and gravitational mass were the same to one part in 1011; this was a more accurate determination of the equivalence principle than results achieved by Isaac Newton and, later, Friedrich Wilhelm Bessel.

Epicycle—in the Ptolemaic universe, a pattern of small circles traced out by a planet at the edge of its "deferent" as it orbited the Earth; this was how the Greeks accounted for the apparent retrograde motions of the planets.

Equivalence principle—the phenomenon first noted by Galileo that bodies falling in a gravitational field fall at the same rate, independent of their weight and composition; Einstein extended the principle to show that gravitation is identical (equivalent) to acceleration.

Escape velocity—the speed at which a body must travel in order to escape a strong gravitational field; rockets fired into orbits around the Earth have calculated escape velocities, as do galaxies at the periphery of galaxy clusters.

Ether (or aether)—a substance whose origins were in the Greek concept of "quintessence," the ether was the medium through which energy and matter moved, something more than a vacuum and less than air; in the late nineteenth century the Michelson-Morley experiment disproved the existence of the ether.

Euclidean geometry—plane geometry developed by the third-century bc Greek mathematician Euclid; in this geometry, parallel lines never meet.

Fermion—a particle with half-integer spin, like protons and electrons, that make up matter.

Field—a physical term describing the forces between massive bodies in gravity and electric charges in electromagnetism; Michael Faraday discovered the concept of field when studying magnetic conductors.

Field equations—differential equations describing the physical properties of interacting massive particles in gravity and electric charges in electromagnetism; Maxwell's equations for electromagnetism and Einstein's equations of gravity are prominent examples in physics.

Fifth force or **"skew" force**—a new force in MOG that has the effect of modifying gravity over limited length scales; it is carried by a particle with mass called the phion.

Fine-tuning—the unnatural cancellation of two or more large numbers involving an absurd number of decimal places, when one is attempting to explain a physical phenomenon; this signals that a true understanding of the physical phenomenon has not been achieved.

Fixed stars—an ancient Greek concept in which all the stars were static in the sky, and moved around the Earth on a distant crystalline sphere.

Frame of reference—the three spatial coordinates and one time coordinate that an observer uses to denote the position of a particle in space and time.

Galaxy—organized group of hundreds of billions of stars, such as our Milky Way.

Galaxy cluster—many galaxies held together by mutual gravity but not in as organized a fashion as stars within a single galaxy.

Galaxy rotation curve—a plot of the Doppler shift data recording the observed velocities of stars in galaxies; those stars at the periphery of giant spiral galaxies are observed to be going faster than they "should be" according to Newton's and Einstein's gravity theories.

General relativity—Einstein's revolutionary gravity theory, created in 1916 from a mathematical generalization of his theory of special relativity; it changed our concept of gravity from Newton's universal force to the warping of the geometry of spacetime in the presence of matter and energy.

Geodesic—the shortest path between two neighboring points, which is a straight line in Euclidian geometry, and a unique curved path in four-dimensional spacetime.

Globular cluster—a relatively small, dense system of up to millions of stars occurring commonly in galaxies.

Gravitational lensing—the bending of light by the curvature of spacetime; galaxies and clusters of galaxies act as lenses, distorting the images of distant bright galaxies or quasars as the light passes through or near them.

Gravitational mass—the active mass of a body that produces a gravitational force on other bodies.

Gravitational waves—ripples in the curvature of spacetime predicted by general relativity; although any accelerating body can produce gravitational radiation or waves, those that could be detected by experiments would be caused by cataclysmic cosmic events.

Graviton—the hypothetical smallest packet of gravitational energy, comparable to the photon for electromagnetic energy; the graviton has not yet been seen experimentally.

Group (in mathematics)—in abstract algebra, a set that obeys a binary operation that satisfies certain axioms; for example, the property of addition of integers makes a group; the branch of mathematics that studies groups is called group theory.

Hadron—a generic word for fermion particles that undergo strong nuclear interactions.

Hamiltonian—an alternative way of deriving the differential equations of motion for a physical system using the calculus of variations; Hamilton's principle is also called the "principle of stationary action" and was originally formulated by Sir William Rowan Hamilton for classical mechanics; the principle applies

to classical fields such as the gravitational and electromagnetic fields, and has had important applications in quantum mechanics and quantum field theory.

Homogeneous—in cosmology, when the universe appears the same to all observers, no matter where they are in the universe.

Inertia—the tendency of a body to remain in uniform motion once it is moving, and to stay at rest if it is at rest; Galileo discovered the law of inertia in the early seventeenth century.

Inertial mass—the mass of a body that resists an external force; since Newton, it has been known experimentally that inertial and gravitational mass are equal; Einstein used this equivalence of inertial and gravitational mass to postulate his equivalence principle, which was a cornerstone of his gravity theory.

Inflation theory—a theory proposed by Alan Guth and others to resolve the flatness, horizon, and homogeneity problems in the standard big bang model; the very early universe is pictured as expanding exponentially fast in a fraction of a second.

Interferometry—the use of two or more telescopes, which in combination create a receiver in effect as large as the distance between them; radio astronomy in particular makes use of interferometry.

Inverse square law—discovered by Newton, based on earlier work by Kepler, this law states that the force of gravity between two massive bodies or point particles decreases as the inverse square of the distance between them.

Isotropic—in cosmology, when the universe looks the same to an observer, no matter in which direction she looks.

Kelvin temperature scale—designed by Lord Kelvin (William Thomson) in the mid-1800s to measure very cold temperatures, its starting point is absolute zero, the coldest possible temperature in the universe, corresponding to −273.15 degrees Celsius; water's freezing point is 273.15K (0°C), while its boiling point is 373.15K (100°C).

Lagrange points—discovered by the Italian-French mathematician Joseph-Louis Lagrange, these five special points are in the vicinity of two orbiting masses where a third, smaller mass can orbit at a fixed distance from the larger masses; at the Lagrange points, the gravitational pull of the two large masses precisely equals the centripetal force required to keep the third body, such as a satellite, in a bound orbit; three of the Lagrange points are unstable, two are stable.

Lagrangian—named after Joseph-Louis Lagrange, and denoted by L, this mathematical expression summarizes the dynamical properties of a physical system; it is defined in classical mechanics as the kinetic energy T minus the potential energy V; the equations of motion of a system of particles may be derived from the Euler-Lagrange equations, a family of partial differential equations.

Light cone—a mathematical means of expressing past, present, and future space and time in terms of spacetime geometry; in four-dimensional Minkowski spacetime, the light rays emanating from or arriving at an event separate spacetime into a past cone and a future cone which meet at a point corresponding

to the event.

Lorentz transformations—

mathematical transformations from one inertial frame of reference to another such that the laws of physics remain the same; named after Hendrik Lorentz, who developed them in 1904, these transformations form the basic mathematical equations underlying special relativity.

Mercury anomaly—a phenomenon in which the perihelion of Mercury's orbit advances more rapidly than predicted by Newton's equations of gravity; when Einstein showed that his gravity theory predicted the anomalous precession, it was the first empirical evidence that general relativity might be correct.

Meson—a short-lived boson composed of a quark and an antiquark, believed to bind protons and neutrons together in the atomic nucleus.

Metric tensor—mathematical symmetric tensor coefficients that determine the infinitesimal distance between two points in spacetime; in effect the metric tensor distinguishes between Euclidean and non-Euclidean geometry.

Michelson-Morley experiment—1887 experiment by Albert Michelson and Edward Morley that proved that the ether did not exist; beams of light traveling in the same direction, and in the perpendicular direction, as the supposed ether showed no difference in speed or arrival time at their destination.

Milky Way—the spiral galaxy that contains our solar system.

Minkowski spacetime—the geometrically flat spacetime, with no gravitational effects, first described by the Swiss mathematician Hermann Minkowski; it became the setting of Einstein's theory of gravity.

MOG—my relativistic modified theory of gravitation, which generalizes Einstein's general relativity; MOG stands for "Modified Gravity."

MOND—a modification of Newtonian gravity published by Mordehai Milgrom in 1983; this is a nonrelativistic phenomenological model used to describe rotational velocity curves of galaxies; MOND stands for "Modified Newtonian

Dynamics."

Neutrino—an elementary particle with zero electric charge; very difficult to detect, it is created in radioactive decays and is able to pass through matter almost undisturbed; it is considered to have a tiny mass that has not yet been accurately measured.

Neutron—an elementary and electrically neutral particle found in the atomic nucleus, and having about the same mass as the proton.

Nuclear force—another name for the strong force that binds protons and neutrons together in the atomic nucleus.

Nucleon—a generic name for a proton or neutron within the atomic nucleus.

Neutron star—the collapsed core of a star that remains after a supernova explosion; it is extremely dense, relatively small, and composed of neutrons.

Newton's gravitational constant—the constant of proportionality, G, which occurs in the Newtonian law of gravitation, and says that the attractive force between

two bodies is proportional to the product of their masses and inversely proportional to the square of the distance between them; its numerical value is: $G = 6.67428 \pm 0.00067 \times 10{-11}\ m3\ kg{-1}\ s{-2}$.

Nonsymmetric field theory (Einstein)—a mathematical description of the geometry of spacetime based on a metric tensor that has both a symmetric part and an antisymmetric part; Einstein used this geometry to formulate a unified field

theory of gravitation and electromagnetism.

Nonsymmetric Gravitation Theory (NGT)—my generalization of Einstein's purely gravitation theory (general relativity) that introduces the antisymmetric field as an extra component of the gravitational field; mathematically speaking, the nonsymmetric field structure is described by a non-Riemannian geometry.

Parallax—the apparent movement of a nearer object relative to a distant background when one views the object from two different positions; used with triangulation for measuring distances in astronomy.

Paradigm shift—a revolutionary change in belief, popularized by the philosopher Thomas Kuhn, in which the majority of scientists in a given field discard a traditional theory of nature in favor of a new one that passes the tests of experiment and observation; Darwin's theory of natural selection, Newton's gravity theory, and Einstein's general relativity all represented paradigm shifts.

Parsec—a unit of astronomical length equal to 3.262 light years.

Particle-wave duality—the fact that light in all parts of the electromagnetic spectrum (including radio waves, X-rays, etc., as well as visible light) sometimes acts like waves and sometimes acts like particles or photons; gravitation may be similar, manifesting as waves in spacetime or graviton particles.

Perihelion—the position in a planet's elliptical orbit when it is closest to the sun.

Perihelion advance—the movement, or changes, in the position of a planet's perihelion in successive revolutions of its orbit over time; the most dramatic perihelion advance is Mercury's, whose orbit traces a rosette pattern.

Perturbation theory—a mathematical method for finding an approximate solution to an equation that cannot be solved exactly, by expanding the solution in a series in which each successive term is smaller than the preceding one.

Phion—name given to the massive vector field in MOG; it is represented both by a boson particle, which carries the fifth force, and a field.

Photoelectric effect—the ejection of electrons from a metal by X-rays, which proved the existence of photons; Einstein's explanation of this effect in 1905 won him the Nobel Prize in 1921; separate experiments proving and demonstrating

the existence of photons were performed in 1922 by Thomas Millikan and Arthur Compton, who received the Nobel Prize for this work in 1923 and 1927, respectively.

Photon—the quantum particle that carries the energy of electromagnetic waves; the spin of the photon is 1 times Planck's constant h.

Pioneer 10 and 11 spacecraft—launched by NASA in the early 1970s to explore the outer solar system, these spacecraft show a small, anomalous acceleration as they leave the inner solar system.

Planck's constant (h)—a fundamental constant that plays a crucial role in quantum mechanics, determining the size of quantum packages of energy such as the photon; it is named after Max Planck, a founder of quantum mechanics

Principle of general covariance—Einstein's principle that the laws of physics remain the same whatever the frame of reference an observer uses to measure physical quantities.

Principle of least action—more accurately the principle of *stationary* action, this variational principle, when applied to a mechanical system or a field theory, can be used to derive the equations of motion of the system; the credit for discovering the principle is given to Pierre-Louis Moreau Maupertius but it may have been discovered independently by Leonhard Euler or Gottfried Leibniz.

Proton—an elementary particle that carries positive electrical charge and is the nucleus of a hydrogen atom.

Ptolemaic model of the universe—the predominant theory of the universe until the Renaissance, in which the Earth was the heavy center of the universe and all other heavenly bodies, including the moon, sun, planets, and stars, orbited around it; named for Claudius Ptolemy.

Quantize—to apply the principles of quantum mechanics to the behavior of matter and energy (such as the electromagnetic or gravitational field energy); breaking down a field into its smallest units or packets of energy.

Quantum field theory—the modern relativistic version of quantum mechanics used to describe the physics of elementary particles; it can also be used in nonrelativistic fieldlike systems in condensed matter physics.

Quantum gravity—attempts to unify gravity with quantum mechanics.

Quantum mechanics—the theory of the interaction between quanta (radiation) and matter; the effects of quantum mechanics become observable at the submicroscopic distance scales of atomic and particle physics, but macroscopic quantum effects can be seen in the phenomenon of quantum entanglement.

Quantum spin—the intrinsic quantum angular momentum of an elementary particle; this is in contrast to the classical orbital angular momentum of a body rotating about a point in space.

Quark—the fundamental constituent of all particles that interact through the strong nuclear force; quarks are fractionally charged, and come in several varieties; because they are confined within particles such as protons and neutrons, they cannot be detected as free particles.

Quasars—"quasi-stellar objects," the farthest distant objects that can be detected with radio and optical telescopes; they are exceedingly bright, and are believed to be young, newly forming galaxies; it was the discovery of quasars in 1960 that quashed the steady-state theory of the universe in favor of the big bang.

Quintessence—a fifth element in the ancient Greek worldview, along with earth, water, fire and air, whose purpose was to move the crystalline spheres that supported the heavenly bodies orbiting the Earth; eventually this concept became known as the "ether," which provided the *something* that bodies needed to be in contact with in order to move; although Einstein's special theory of relativity dispensed with the ether, recent explanations of the acceleration of the universe call the varying negative pressure vacuum energy "quintessence."

Redshift—a useful phenomenon based on the Doppler principle that can indicate whether and how fast bodies in the universe are receding from an observer's position on Earth; as galaxies move rapidly away from us, the frequency of the wavelength of their light is shifted toward the red end of the electromagnetic

spectrum; the amount of this shifting indicates the distance of the galaxy.

Riemann curvature tensor—a mathematical term that specifies the curvature of four-dimensional spacetime.

Riemannian geometry—a non-Euclidean geometry developed in the mid-nineteenth century by the German mathematician George Bernhard Riemann that describes curved surfaces on which parallel lines *can* converge, diverge, and even intersect, unlike Euclidean geometry; Einstein made Riemannian geometry the mathematical formalism of general relativity.

Satellite galaxy—a galaxy that orbits a host galaxy or even a cluster of galaxies.

Scalar field—a physical term that associates a value without direction to every point in space, such as temperature, density, and pressure; this is in contrast to a vector field, which has a direction in space; in Newtonian physics or in electrostatics, the potential energy is a scalar field and its gradient is the vector force field; in quantum field theory, a scalar field describes a boson particle with spin zero.

Scale invariance—distribution of objects or patterns such that the same shapes and distributions remain if one increases or decreases the size of the length scales or space in which the objects are observed; a common example of scale invariance

is fractal patterns.

Schwarzschild solution—an exact spherically symmetric static solution of Einstein's field equations in general relativity, worked out by the astronomer Karl Schwarzschild in 1916, which predicted the existence of black holes.

Self-gravitating system—a group of objects or astrophysical bodies held together by mutual gravitation, such as a cluster of galaxies; this is in contrast to a "bound system" like our solar system, in which bodies are mainly attracted to and revolve around a central mass.

Singularity—a place where the solutions of differential equations break down; a spacetime singularity is a position in space where quantities used to determine the gravitational field become infinite; such quantities include the curvature of spacetime and the density of matter.

Spacetime—in relativity theory, a combination of the three dimensions of space with time into a four-dimensional geometry; first introduced into relativity by Hermann Minkowski in 1908.

Special theory of relativity—Einstein's initial theory of relativity, published in 1905, in which he explored the "special" case of transforming the laws of physics from one uniformly moving frame of reference to another; the equations

of special relativity revealed that the speed of light is a constant, that objects appear contracted in the direction of motion when moving at close to the speed of light, and that $E = mc2$, or energy is equal to mass times the speed of light squared.

Spin—see quantum spin.

String theory—a theory based on the idea that the smallest units of matter are not point particles but vibrating strings; a popular research pursuit in physics for two decades, string theory has some attractive mathematical features, but has yet to make a testable prediction.

Strong force—see nuclear force.

Supernova—spectacular, brilliant death of a star by explosion and the release of heavy elements into space; supernovae type 1a are assumed to have the same intrinsic brightness and are therefore used as standard candles in estimating cosmic distances.

Supersymmetry—a theory developed in the 1970s which, proponents claim, describes the most fundamental spacetime symmetry of particle physics: For every boson particle there is a supersymmetric fermion partner, and for every fermion there exists a supersymmetric boson partner; to date, no supersymmetric particle partner has been detected.

Tully-Fisher law—a relation stating that the asymptotically flat rotational velocity of a star in a galaxy, raised to the fourth power, is proportional to the mass or luminosity of the galaxy.

Unified theory (or unified field theory)—a theory that unites the forces of nature; in Einstein's day those forces consisted of electromagnetism and gravity; today the weak and strong nuclear forces must also be taken into account, and perhaps someday MOG's fifth force or skew force will be included; no one has yet discovered a successful unified theory.

Vacuum—in quantum mechanics, the lowest energy state, which corresponds to the vacuum state of particle physics; the vacuum in modern quantum field theory is the state of perfect balance of creation and annihilation of particles and antiparticles.

Variable Speed of Light (VSL) cosmology—an alternative to inflation theory, in which the speed of light was much faster at the beginning of the universe than it is today; like inflation, this theory solves the horizon and flatness problems in the very early universe in the standard big bang model.

Vector field—a physical value that assigns a field with the position and direction of a vector in space; it describes the force field of gravity or the electric and magnetic force fields in James Clerk Maxwell's field equations.

Virial theorem—a means of estimating the average speed of galaxies within galaxy clusters from their estimated average kinetic and potential energies.

Vulcan—a hypothetical planet predicted by the nineteenth-century astronomer Urbain Jean Joseph Le Verrier to be the closest orbiting planet to the sun; the presence of Vulcan would explain the anomalous precession of the perihelion of Mercury's orbit; Einstein later explained the anomalous precession in general relativity by gravity alone.

Weak force—one of the four fundamental forces of nature, associated with radioactivity such as beta decay in subatomic physics; it is much weaker than the strong nuclear force but still much stronger than gravity.

X-ray clusters—galaxy clusters with large amounts of extremely hot gas within them that emit X-rays; in such clusters, this hot gas represents at least twice the mass of the luminous stars.

About the Author

Balungi Francis was born in Kampala, Uganda, to a single poor mother, grew up in Kawempe, and later joined Makerere Universty in 2006, graduating with a Bachelor Science degree in Land Surveying in 2010. For four years he taught in Kampala City high schools, majoring in the fields of Gravitation and Quantum Physics. His first book, "Mathematical Foundation of the Quantum theory of Gravity," won the Young Kampala Innovative Prize and was mentioned in the African Next Einstein Book Prize (ANE).

He has spent over 15years researching and discovering connections in physics, mathematics, geometry, cosmology, quantum mechanics, gravity, in addition to astrophysics, unified physics and geographical information systems . These studies led to his groundbreaking theories, published papers, books and patented inventions in the science of Quantum Gravity, which have received worldwide recognition.

From these discoveries, Balungi founded the SUSP (Solutions to the Unsolved Scientific Problems) Project Foundation in 2004 - now known as the SUSP Science Foundation. As its current Director of Research, Balungi leads physicists, mathematicians and engineers in exploring Quantum Gravity principles and their implications in our world today and for future generations.

Balungi launched the Visionary School of Quantum Gravity in 2016 in order to bring the learning and community further together. It's the first and only Quantum Gravity physics program of its kind, educating thousands of students from over 80 countries.

The book "Quantum Gravity in a Nutshell1", a most recommend book in quantum gravity research, was produced based on Balungi's discoveries and their potential for generations to come. Balungi is currently guiding the Foundation, speaking to audiences worldwide, and continuing his groundbreaking research.

Email:

bfrancis@cedat.mak.ac.ug, balungif@gmail.com